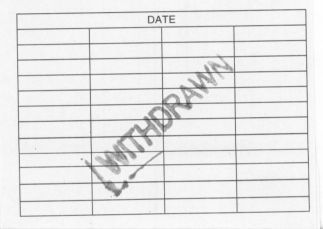

Evolution at a Crossroads

⅃Ɫ   Bradford Books

Elliott Sober, Editor. CONCEPTUAL ISSUES IN EVOLUTIONARY BIOL-
OGY. 1984.

Robert N. Brandon and Richard M. Burian, Editors. GENES, ORGANISMS,
POPULATIONS. 1984.

Elliott Sober. THE NATURE OF SELECTION. 1984.

David J. Depew and Bruce H. Weber, Editors. EVOLUTION AT A CROSS-
ROADS. 1985.

Evolution at a Crossroads: The New Biology and the New Philosophy of Science

edited by
David J. Depew and
Bruce H. Weber

A Bradford Book
The MIT Press
Cambridge, Massachusetts
London, England

Second printing, 1986

© 1985 by
The Massachusetts Institute of Technology

This book was set in Palatino
by The MIT Press Computergraphics Department
and printed and bound by The Murray Printing Co.
in the United States of America.

Library of Congress Cataloging in Publication Data

Main entry under title:
Evolution at a crossroads.

"A Bradford book."
Bibliography: p.
Includes index.
1. Biology—Philosophy. 2. Evolution—Philosophy.
I. Depew, David J., 1942–   . II. Weber, Bruce H.
QH331.E86 1985   574'.01   84–21829
ISBN 0–262–04079–4

Contents

Acknowledgments

More than half of the essays appearing in this volume were prepared for, or in response to, the Thirteenth Annual Philosophy Symposium, held at California State University, Fullerton, for a week in the Spring of 1983. The editors wish to thank the following people for help in arranging and financing that symposium: Jewel Plummer Cobb, president of California State University, Fullerton; Don Schweitzer, dean of the School of Humanities and Social Sciences; the members of the CSUF Department of Philosophy and the CSUF Institute for Molecular Biology; the Philosophy Club; and the CSUF Department Associations Council.

The remaining papers were solicited from scholars working on the frontier where philosophy meets evolutionary and developmental biology. None of the essays has been previously published. The editors are grateful to the authors for their prompt and generous cooperation in the editing process. Special thanks are due to Marjorie Grene and Richard Burian, and to Betty Stanton of Bradford Books, for helping in various ways to give birth to the book. The late Mark Lackey helped with the references, and Gordon Brown's word-processing skill and editorial sensitivity made him a valued collaborator. The editors also gratefully acknowledge a grant from the CSUF Faculty Research Committee for preparation of the manuscript. The secretarial staffs of the CSUF departments of philosophy and chemistry provided useful assistance of all sorts.

Introduction

For half a century the Modern Evolutionary Synthesis, or neo-Darwinism, has provided a virtually unchallenged theoretical framework within which biological inquiry as a whole has worked. Even while biologists have devoted most of their efforts to unraveling detailed mysteries of organic structure and function, they have confidently presupposed that neo-Darwinism offers a secure and elegant set of principles that explain why the structures they investigate have arisen and have come to work the way they do.

Neo-Darwinism married classical Darwinian adaptationism to rediscovered Mendelian genetics by postulating that randomly generated genetic variations could be amplified and established within interbreeding populations by forces acting to preserve the adaptively valuable consequences of genetic variation. Evolution was seen as a long-range function of these gradual changes in gene frequencies.

Accordingly, the makers of the Modern Synthesis were inclined to congratulate Darwin for remaining studiously neutral about the source of this variation, so that in the fullness of time, when Mendelian genetics arrived on the scene, it could be fitted unambiguously into the limited position ostensibly left for it by Darwin. We need not judge here how far this account accurately reflects Darwin's views. To say the least, this reconstruction is decidedly Whiggish. Rather, our concern is to note that the synthesists were equally prone to congratulate themselves for staying judiciously neutral about the detailed physical structure and working of the gene. For they confidently assumed that in another fullness of time, when this knowledge became available, it would overturn no part of their work.

Things have generally gone well, for molecular genetics has given us precise knowledge of the relevant mechanisms of heredity and variation. Within the last decade, however, neo-Darwinism has come under a variety of pressures from advanced molecular biology. In the process certain background assumptions associated with the received neo-Darwinian paradigm have been called into question.

First, selection can plausibly be thought of as occurring at a number of different levels and on a number of different sorts of objects, from genes to demes to species. This threatens the exclusive focus on competition among individual organisms that is characteristic of Darwinian thinking (cf. Brandon and Burian, 1984, for a collection of the most important literature). Second, gene frequencies can be fixed without selection pressure by such processes as drift and drive (Kimura, 1979; Dover, 1982). This undermines the pan-adaptationist expectations of some Darwinians. Third, periods of rapid evolutionary change can be followed by periods of stability. This puts in doubt the gradualist assumptions built into Darwinian and neo-Darwinian models (Eldredge and Gould, 1972; Gould, 1980, 1982a,b). Finally, appreciation of the complex structure of the genome fails to assure us that genes are as mutually independent or as passive in the evolutionary process as traditional Darwinian anti-Lamarckism has assumed (Hunkapiller et al., in Milkman, 1982).

It is small wonder that in this atmosphere an intense debate has arisen about whether neo-Darwinism can be re-articulated in a way that embraces our current and prospective knowledge of genetic change; or whether, on the contrary, a coherent grasp of the facts and a steady way into the future can be attained only by breaking with neo-Darwinian patterns of conceptualization generally. Of particular concern is how far two working principles that have thus far guided cooperation between molecular biology and modern evolutionary theory remain useful. The first of these principles is the Hardy-Weinberg Equilibrium, which generalizes Mendel's Laws to state how genes will be distributed in an interbreeding population not subject to selection or other disturbances. The second is the so-called Central Dogma of classical molecular biology. This states that information flows exclusively from, and never to, DNA. A new appreciation of evolutionary complexity renders it increasingly doubtful that these principles alone provide a sufficiently solid foundation on which to build a comprehensive evolutionary theory.

The authors of the chapters in this book are eminent biologists who have shown themselves more than casually sensitive to the conceptual side of their work and philosophers of science in possession of more than a passing knowledge of contemporary problems in biology. Their essays vary widely in topic and approach and so illustrate the wealth and diversity of concerns that today sustain a lively dialogue between philosophers and evolutionary biologists. The essays fall roughly into three conceptual domains. Some explore topics about rationality and progress in science generally, and in biology in particular. Others focus on the troubled relation between biology and reductionistic programs in the philosophy of science. Still others are concerned with the relative roles of chance, law, and purposiveness in biological systems, including the genome itself. The intricate pattern of agreement and disagreement among the essays can most readily be brought to light when we note that the authors generally treat these topics with an eye on the problems and prospects of contemporary evolutionary theory.

First, the philosophers who have contributed to this collection are aware that continued harmony between molecular biology and the Darwinian tradition depends in part on how we are to think generally about continuity and discontinuity within and between scientific theories. As it happens, at the same time that the fate of neo-Darwinism has begun to concern biologists, philosophers of science have been engaged in a sustained and illuminating discussion about scientific change and progress. Indeed, talk about a breakdown in the Darwinian tradition has been formulated in an atmosphere in which scientists and philosophers, and even the public, have been quick to pick up the scent of scientific revolutions and paradigm crises, just as our earlier confidence in neo-Darwinism was doubtless affected by an optimism about the cumulative rationality of science that was widespread in the culture during the heyday of the synthesis. Perhaps under these circumstances the implications of recent difficulties in evolutionary biology have been exaggerated as much as they were previously minimized. In the wake of Kuhn's work it at first appeared that philosophers might be forced to maintain a flat-footed commitment to a rigidly normalized paradigm as the only alternative to a revolt that must end in mutual incomprehension and fragmentation. It is particularly important to note, then, that a consensus is now emerging among philosophers that this is not the case.

The terms "new" and "post-Kuhnian" philosophy of science are rubrics under which are assembled many different and often conflicting insights and suggestions. All of these, however, hold out the possibility that in the development of science there can be deep continuity through and across much change, and mutual comprehension even where there is much disagreement. Marjorie Grene's essay, in which the themes of post-Kuhnian philosophy of science are articulated from the perspective of a biologically oriented philosopher, not only makes a contribution to the further development of this new approach but provides the best preface to the other essays in the collection. In general, by casting the vicissitudes of contemporary evolutionary biology in this light, the philosophers who have contributed to this book provide a background against which the prospective transformation of evolutionary biology can be profitably viewed, and perhaps even carried out, without demanding too much overt continuity or being scandalized at its occasional absence.

Second, whether the neo-Darwinian paradigm can, as Stephen Jay Gould has put it, be "expanded" (Gould, 1982b) to accommodate selection at various levels, at different rates, and to find room for some nonselectionist processes, depends not a little on the things to which that paradigm commits us in the first place. It has recently been acknowledged that the synthesis was, in its earliest phases (particularly in the version sponsored by Sewall Wright), more pluralistic than its later canonical formulations (Mayr and Provine, 1977; Provine, 1983; Gould in Grene, 1983). That neo-Darwinians soon began to downplay this pluralism may in part be traced to entanglement of the synthesis with reductionistic programs in the philosophy of science, which were widely pursued at the time of its diffusion.

Francisco Ayala and Ernst Mayr suggest, however, that even in its canonical formulations neo-Darwinism is not, or need not have been, as reductionistic as its opponents and some of its supporters have alleged. The possibility that new developments will lead to a more general rearticulation of neo-Darwinian insights is accordingly increased by the more supple reconstructions of its scope and methods suggested by Mayr and explored by a number of our contributors. Such explorations can be seen in the arguments deployed by Robert Brandon and C. Dyke against the reductionistically inspired programs of Richard Dawkins and some sociobiologists. Of particular importance is Dyke's claim that interpretive methods previously confined to social science are of some relevance to evolutionary biology, once the demand for reduction has been alleviated and the insights of post-Kuhnian philosophy of science have been absorbed. Indeed, Gunther Stent argues in his essay that such methods are indispensable to developmental biology.

Finally, if we are accurately to judge whether our new and prospective knowledge of the complex structure and function of the genome can be accommodated by an appropriately expanded neo-Darwinism, we shall have to decide what conceptual matrix most perspicuously represents that knowledge. We shall then have to estimate how the result squares with our neo-Darwinian inheritance. There is little doubt that this is frontier country in evolutionary biology. Several contributors to this volume argue that there are endogenous factors within the genome that have evolutionary roles and relevance apart from selection as classically understood. For John Campbell, understanding the evolutionary function of these internal factors will require redeployment of philosophical conceptions of purposiveness explicitly rejected by the Darwinian tradition at its inception. Stuart A. Kauffman argues, alternatively, that understanding these factors will lead us to a level of ahistorical universality and statistical lawfulness lying well below the historical *nexus* of chance and *telos* familiar to Darwinians. But if our growing appreciation of the internal complexity of the genome can be accommodated only by conceptions such as those proposed by Campbell or Kauffman, it becomes a matter of the highest importance to estimate whether and how the neo-Darwinian inheritance is to be projected into the future. The philosophical framework within which these essays are set makes us ready to believe, however, that the various suggestions of our authors are already contributing to a future for evolutionary theory that is interestingly continuous with its distinguished past.

This is perhaps introduction enough to essays and essayists who speak well for themselves. We have appended an interpretive essay in which interrelationships among the various contributions are further explored. We suggest more there about how aspects of recent philosophy of science can be brought to bear on problems concerning expansion and continuity in evolutionary theory.

References

Brandon, R., and R. Burian, 1984. *Genes, Organisms, Populations: Controversies over the Units of Selection*, Cambridge, Massachusetts: The MIT Press. A Bradford book.

Dover, G., 1982. Molecular drive: a cohesive mode of species evolution. *Nature* 299:111–117.

Eldredge, N., and S. J. Gould, 1972. Punctuated equilibria: an alternative to phyletic gradualism. In *Models in Paleobiology*, T. J. M. Schoff, ed. San Francisco: W. H. Freeman, pp. 82–115.

Gould, S. J., 1980. Is a new general theory of evolution emerging? *Paleobiology* 6:119–130.

Gould, S. J., 1982a. The meaning of punctuated equilibrium and its role in validating a hierarchical approach to macroevolution. In *Perspectives in Evolution*, R. Milkman, ed. Sunderland, Massachusetts: Sinauer Assoc., pp. 83–104.

Gould, S. J., 1982b. Change in developmental tuning as a mechanism of macroevolution. In *Evolution and Development*, J. T. Bonner, ed. Heidelberg: Springer Verlag, pp. 333–346.

Gould, S. J., 1983. The hardening of the modern synthesis. In *Dimensions of Darwinism: Themes and Counterthemes in Twentieth-Century Evolutionary Theory*, M. Grene, ed. Cambridge: Cambridge University Press, pp. 71–93.

Grene, M., ed., 1983. *Dimensions of Darwinism: Themes and Counterthemes in Twentieth-Century Evolutionary Theory*. Cambridge: Cambridge University Press.

Hunkapiller, T., H. Huang, L. Hood, and J. H. Campbell, 1982. The impact of modern genetics on evolutionary theory. In *Perspectives in Evolution*, R. Milkman, ed. Sunderland, Massachusetts: Sinauer Assoc., pp. 164–189.

Kimura, M., 1979. The neutral theory of molecular evolution. *Scientific American* 241(5):98–126.

Mayr, E., and W. B. Provine, eds., 1980. *The Evolutionary Synthesis*. Cambridge, Massachusetts: Harvard University Press.

Provine, W. B., 1983. The development of Wright's theory of evolution: systematics, adaptation, and drift. In *Dimensions of Darwiniswm: Themes and counterthemes in Twentieth-Century Evolutionary Theory*, M. Grene, ed. Cambridge: Cambridge University Press, pp. 43–70.

Evolution at a Crossroads

Perception, Interpretation, and the Sciences: Toward a New Philosophy of Science

Marjorie Grene

Current issues in the philosophy of biology illustrate, and are most fruitfully approached, in terms of a philosophical perspective less monolithic than the once regnant logical empiricism but less revolutionary than the adherents of what was taken to be Thomas Kuhn's position. My aim is to sketch such a "new philosophy of science," a title borrowed from Harold Brown (Brown, 1977), in order to provide a general interpretive framework for this book. Not all our authors, of course, subscribe to this new philosophy. Mayr, Stent, and Dyke clearly seem to do so; Burian, Campbell, Brandon, and perhaps Kauffman might, by implication. But even a more conservative thinker like Ayala can be well understood within this pluralistic, historically based approach.

There are several reasons a philosophy of biology should fit into the frame presented here. First, biology is more obviously realistic than more abstract sciences. It concerns aspects of living things, not floating phenomena. Even dry bones were once alive. Second, biologists (evolutionary biologists, certainly) know that they themselves are animals, trying, like other animals, to orient themselves to their peculiar environments, whether in field observation, laboratory experiment, or theory. Third, biology, and especially evolutionary biology, more than most fields, is sensitive to its own history—to the "growth of biological thought," to quote the title of Mayr's recent magnum opus (Mayr, 1982). Current controversies in evolution, for example, concern not only the structure, but also the history of the evolutionary synthesis (Gould, 1983; Provine, 1983). Fourth, biological disciplines have already proved strikingly suitable for a pluralistic approach to questions of the interrelations among the sciences. (See the work on interfield theories by Darden and Maull, 1977; Maull, 1977; and Darden, 1983.) For all these reasons I hope that what follows will serve to elicit the underlying unity among the closely related yet multifarious contributions to this book.

The phrase "new philosophy of science" by now suggests a turn from logical reconstruction to history, and that is indeed part of what is intended here. Logical empiricism, the older orthodoxy in philosophy of science, treated science as a logical edifice in which observation statements (the ground floor) were generalized into laws, which in turn were explained by theories (the top story). There was supposed to be a deductive relation between theories and laws, and a predictive relation between theories and the lower levels. It was all a very formal, abstract affair, disclaiming any interest in science as a process, as a developing segment of human history. That is now changing. A chart contrasting the approach of the new perspective with the old would include the following points:

New	Old
• Primacy of perception	• Phenomenalism (sensation/judgment)
• Primacy of orientation	• Hypothetico-deductivism or inductivism
• Comprehensive realism	• Positivism or thin realism
• Primacy of history	• Discovery/justification
• Progressive problem-solving	• Ahistorical (linear)/incommensurability
• Inclusiveness of hermeneutics	• Nomothetic/idiographic "science"
• Sciences as forms of life:	• Fact/value:
values *in* science	objective/subjective
• Social nature of science:	• Science outside society:
1. Sciences as communities	1. Irrelevant
2. Learning (tradition) in science	2. Irrelevant
3. Science in society	3. Science prescriptive for society
• Scientific pluralism	• Unity of science

I will use these points as an outline. Before I begin, however, one general introductory remark is needed. The best-known versions of the turn to history connote a relativism and subjectivism, or at best a pragmatism, that I am concerned to avoid. Following the lead of John Compton, I prefer to start with what seems to me an even more fundamental move: the change from a phenomenalist or positivist starting point for philosophy of science to a new type of realism, a realism that is primary, not inferred (Compton, 1983). Scientists are real people trying, like other creatures, to solve problems presented to them by some aspect of the world that specially concerns them. As Heidegger put it, the primary structure of *Dasein* (human being)—but also, I believe, of the being of all animals—is care (Heidegger, 1927, pp. 199ff.). To insist that scientists are real people facing real problems does not mean in this context that they are concerned with practical problems—how to cure cancer or clean up pollution; nor does world here mean environment in the environmentalist's sense—oil slicks or acid rain. The problems of scientists are questions about how something in the real world really works, whether it be the reproductive system of the aardvark, the origin of galaxies, or the production of a protein by a particular gene or gene complex. Accordingly, the world of the scientist is the field or the laboratory. As one biologist has put it, science begins with what is going on in the lab next door. There is a problem about some real process, however refined and esoteric its formulation, that my colleagues and I are puzzled by in this minicultural, but real, situation. That is the starting point of any scientific inquiry.

However strange the vocabulary and procedures of these inquiries, they are only extensions and modifications of the kinds of inquiries all of us, and even all animals, make about the world around us. The oddity of science is that it is a specifically theoretical inquiry about what is actually the case, not about what is useful or just aesthetically appealing. It is the attempt to orient oneself in one's own scientific environment that constitutes the scientist's activity as scientist. Such activity is an exploration whose structure forms an extrapolation from, or elaboration of, the fundamental structures and processes of ordinary perception.

The Primacy of Perception

Scientific investigations and extrascientific explorations alike begin with the use of the organism's perceptual systems, move to higher levels of abstraction through processes analogous to perception, and return to perception, or instrument-mediated extensions of perception, for their substantiation. To develop such a view, however, one needs a theory of perception radically different from the traditional one. Basically, the older philosophy of science was positivistic (or phenomenalistic). It held that science, like all human awareness, is built up inferentially from least sensory units. Thus, for example, we were supposed to "see" (as colored bits on the retina) spots of, say, black and white, and from this to infer by some long-established and by now unconscious habit a reading, say, on a thermometer. Even if arguments were given from that unlikely starting point, leading to the assertion that there is a reality beyond our consciousness (arguments for a "scientific realism"), this *thin realism*, as I call it, seemed to claim for the external world a kind of hypothetical reality wholly divided, ontologically as well as epistemologically, from the givens on which scientific explanation, like everyday perception, had to be built. Perceptions are hypotheses, says Richard Gregory; and hypotheses, of course, are, on this view, the lifeblood of science (Gregory, 1966). But both perception and hypothetico-deductive science float periously above a sea of mere phenomena, beyond or beneath which one may in turn possibly infer some terra firma, at best a Lockean I-know-not-what.

We now have an alternative account of perception, more plausible in itself, that offers a much more solid foundation for an analysis of scientific discovery and knowledge, namely, J. J. Gibson's ecological theory of perception. Gibson's theory is largely convergent with a philosophical account better known to some philosophers, even to a few philosophers of science, developed by Merleau-Ponty in his major work. However, the *Phenomenology of Perception* (Merleau-Ponty, 1945), together with the earlier *Structure of Behavior* (1942), was necessarily based on now outdated psychological research. I believe, therefore, that we can say more effectively what Merleau-Ponty wanted to say in Gibson's terms.

Gibson's last book is entitled *The Ecological Approach to Visual Perception* (J. J. Gibson, 1979a). His theory, developed during thirty years of experimental work, takes perceivers to be, as they surely are, animals using their perceptual systems to pick up information made available to them from things and events in their environments. This approach is radically opposed to the traditional image-cue model (for vision: retinal image + judgment = perception as hypothesis). It argues, for good reasons based on studies of event perception, the perception of surfaces, distance, and so on, that perceivers take up, relatively directly, information about objects and occurrences available around them insofar as their perceptual systems are adapted to such information pickup, and insofar as their world is so organized as to facilitate it.

Thus the ecological theory distinguishes (1) objects and events in the perceiver's environment; (2) the information specified for the perceiver by features of the ambient array (optic or acoustic); and (3) what the organism perceives by means of such information. First, the environment of any organism is highly structured. It has a definite layout, including objects, occurrences, and in particular, affordances—real features of the world that have a bearing on the organism's way of being in its world. Thus the rabbit affords the hunter a chance of eating, while the hunter affords the rabbit the possibility of being eaten. Such affordances, it must be stressed again, are objective features of what there is. Second, perceptual systems have developed in such a way as to pick up in the ambient array—optic, haptic, acoustic, or, as often happens, intermodal—information systematically available to them. Such an array flows not only because things change and events happen but also because the observer, too, is usually moving. As E. J. Gibson, writing of vision, has recently put it: "Information must be extracted from a flowing optic array. It is the invariants over time in the structure of ambient light registered by the eye-head-brain-body system, that can inform an animal about its environment" (E. J. Gibson, 1984, p. 7).

Third, what the perceiver sees, hears, feels, etc., is not directly the information itself, which is potentially infinite. What he (she) detects, to quote the same text, "are the events and places and things that information specifies, and typically some affordances of them" (E. J. Gibson, 1984, p. 9).

There are corollaries of this view that need to be noted even in this crude sketch. (1) Perceiving entails awareness not only of the object of perception but also of the location of the perceiver's body in his environment: perception includes proprioception. (2) Perception, though direct, is not purely passive. Attention, indeed exploratory activity, is essential to it. It is notorious that we fail to perceive what we fail to attend to, even though much that we do perceive may be marginal to the focus of perception, as our awareness of our own body generally is. (3) Perception can be in error. In contrast to the traditional view, this theory holds that perception is aimed relatively directly at reality, but may miss the mark. The process of perceiving, therefore, includes reality testing—turning one's head, for instance, to see if the boat has really gone downstream, or looking again to see if it is really a bird on the branch, or listening again to find out if the sound really came from there, and so on. (4) Perception, finally and primarily, entails learning, but perceptual learning, not inference added to sensation, as on the traditional view. So, as Gibson notes, the Kantian split between *Anschauung* and *Begriff* must be denied, or at least radically reinterpreted (J. J. Gibson, 1979b). In short, this view of the foundations of science entails Merleau-Ponty's principle of the primacy of perception, against the older image-cue, sensation/judgment theory.

The Primacy of Orientation

A second point of departure may be entitled the primacy of orientation. Every organism, within the limits of its bodily and behavioral endowments, is constantly using the powers available to it to orient itself to its environment, or to some aspect of it. Searching is fundamental to being a live animal. One may contrast, therefore, this ecological, orientational view of perception, and a fortiori of all our knowledge of the world around us, with the anti-biological, anti-ecological hypothetico-deductivism of the older philosophy of science, in which science consists not in seeking (or even finding) but solely in abstractly ordering hypotheses, inferences, and observational conclusions in the right atemporal, abstract sequence. Nor, in my view, was a classic inductivist program any less ahistorical in nature. Starting from Gibson's view of perception, we see by contrast that temporal processes—looking around, searching for clues, finding solutions—are intrinsically linked in the complex process of orientation.

Comprehensive Realism

If we start with Gibsonian perception, we find ourselves in all our activities, including the lifelong labors of scientists, real animals in a real environment trying to solve some problem about a puzzling feature of that environment. Even the abstract, culture-carried environment of the laboratory, after all, is real. Far from being phenomenalists, or would-be realists who have to devise obscure, often science-fictional arguments for our realism, we are realists about science from the start. This is the position of a comprehensive realism, opposed equally to the phenomenalism and to the "thin realism" of the older philosophy of science (Grene, 1983).

It is this position that must be underlined as central to our new conception of science. Neither positivism, which puts the scientist somewhere outside reality, nor classical scientific realism, which tries to argue for the hypothesis that he (she) might be dealing with something other than sensations or conventions, does justice to the aims, procedures, or achievements of scientists and of the sciences. On the other hand, given our twin principles of the primacy of perception and the primacy of orientation, we may affirm realism as a necessary, immediate, and primary ground of scientific inquiry and of scientific knowledge. The scientist's aim is knowledge of the real world because he or she has never left that world. However technical the problems, however esoteric the discourse, however abstruse the equations, they are problems, discourse, equations developed within a subworld of the natural world, a subworld constituted by culture within nature and designed to give us new access to some features of nature itself. If a scientific inquiry becomes so refined as to be dealing with artifact, it has failed as science.

So far as I know, there is as yet no fully articulated expression of the comprehensive realism I am proposing. Recent work by John Compton points in this direction, and the hermeneutical realism of Patrick Heelan is probably convergent with the view adumbrated here (Compton, 1983; Heelan, 1983). However the position may ultimately be articulated, it is amply clear that it is in this direction that work in philosophy of science, both in general and in particular case studies, needs to proceed.

The Primacy of History

Further, the primacy of orientation in the human context entails yet another aspect of our new starting point—the primacy of history. For the process of orientation, in the human case, is contained not only within the natural world in which every animal seeks to find its way but also within the cultural world in which, as human beings, we become the persons we are or strive to be. As I have stated elsewhere, a human person is a personification of nature through participation in, and as an expression of, a culture (Grene, 1974). We are, as Herder has said, invalids of our higher powers. We cannot just naturally be human, as earthworms are naturally earthworms or frogs naturally frogs. To fulfill our capacities, we need, within the constraints set by our genetic endowments and our natural and cultural environments, the medium of a culture, a language, social rituals, the discipline of social practices, through participation in which we develop the human histories that constitute our lives.

This holds a fortiori for the life of science. Traditionally, philosophy of science has formally turned its back on this obvious insight. Philosophy of science used quite self-consciously to insist on what it called the separation of two contexts, the context of discovery and the context of justification. Discovery, it was stressed, had nothing to do with the matter. The chemist Kekulé may have sat before the fire imagining snakes chasing their tails and so may have dreamed up the benzene ring, but that is only psychology, not science. True, philosophy of science should not indeed chase down every psychological or sociological move in the genesis of a scientific idea or a scientific theory. On the other hand, science as nonsearch, science without research, does not and cannot exist. What we need to do is to ask philosophical questions of science in, not apart from, the context of its historical existence. Far from separating the context of discovery from that of justification, therefore, we insist on an internal relation between them. Science as we know it began in the sixteenth and seventeenth centuries in Western Europe. If the drive to discovery, if scientific curiosity about nature's secrets, ever ceases, science may someday simply die off again. Within the framework of its historical existence, however, philosophers of science want to ask what sort of cognitive claims science makes, by what standards it initiates, carries through, and justifies them. How does a discovery differ from a nondiscovery? That is a central question for philosophers of science when they leave their ivory towers and examine the scientific endeavors and achievements of the past 350 years, with the aim of trying to understand what science, as a human cognitive activity, amounts to. Their task is analogous to that of historians of art, who try to analyze a style or a movement, which necessarily comes to be, and *is*, in given historical circumstances, but nevertheless makes its claims as a style or a movement in its own right. Alois Riegl spoke of the *Kunstwollen*, the will to art, of a society (Riegl, 1901; cf. Zerner, 1976). Perhaps all cultures also possess, at least some do, a *Wissenschaftswollen*, a will to science, which exists within given historical constraints yet makes its cognitive claims—claims to truth in other words—within the limits imposed by history.

Once more, this does not make science illusory, arbitrary, or merely pragmatic. The turn to history is not a turn to relativism in either the sense of Feyerabend or in the sense in which Kuhn's doctrine is usually understood. I shall say more about that later. But this warning must be kept in mind. It was in fact to avoid that misunderstanding that I put perception, orientation, and comprehensive realism ahead of the primacy of history in my list of features of the new perspective.

Progressive Problem-Solving

It follows from comprehensive realism and the primacy of history that science consists in progressive problem-solving.

In the older orthodoxy, science was sometimes considered wholly ahistorical. The scientific method was analyzed as quasi-eternal. It may not have happened to happen until the seventeenth century, but it is really always the same. In positivistic terms, there is a linear accumulation of "facts" by means of the "scientific method." But when it became clear, thanks to Kuhn and others, that science comes in different patterns, subject to different "paradigms," as they were called, it was held that because the nonhistorical model had broken down there was simply an incommensurability between one style of science and another. Newton was as good as Einstein, Aristotle as good as Harvey. The obvious fact of the accumulation of scientific knowledge had simply to be denied. Within the new perspective, however, we can see how science both changes conceptual frameworks and leads to an accumulation of knowledge unaffected by such conceptual change. There is both discontinuity in theory, given the alterations of meaning and reference inherent in the growth of any language, including that of science, and continuity, given the contact with reality that scientific activity aims at and often achieves. Priestley can be celebrated in his native Yorkshire as the discoverer of oxygen, even though he went to his grave denying its existence. Richard Owen did discover one of the major evidences for evolution—the homology of the vertebrate limb— although he denied evolution and could even take his discovery as evidence against it. Within the real world, all of us, scientists and nonscientists alike, try to orient ourselves through our instruments of perception, theory, calculation, and so on. When we have succeeded, our achievements may be carried over into a new reading, a conceptual frame, which we ourselves, being who we were in our time, could not have understood. Thus we have in the history of scientific knowledge a peculiar blend of continuity and discontinuity that is precisely one of the distinguishing features of this "peculiar institution."

The Inclusiveness of Hermeneutics

Nevertheless, science is not just a flat statement of facts; it always entails interpretation. Therefore I assert the pervasiveness of interpretation, or, to use more modish language, the inclusiveness of hermeneutics. Science is a style, or even a shifting series of styles, for attempting, often successfully, to interpret reality. It is a question, as people liked to say in the seventeenth century, of reading the text of the world. In the older view natural science was held to be wholly objective. Its aim was to formulate universal laws, from which observational truths could be deduced. In contrast, history or the "human sciences" entailed an ineliminable interpretive factor, from which natural science was happily exempt. Thus a distinction of kind was made between "nomothetic" and "idiographic" disciplines. Even as recently as Charles Taylor's authoritative essay on "Interpretation and the Sciences of Man" (Taylor, 1971), the natural sciences were specifically exempted from this analysis. That was an error. In different degrees and even at different logical levels, there is an interpretive ingredient in all knowledge, scientific as well as historical, in the natural as well as the social sciences.

This thesis was developed in Michael Polanyi's *Personal Knowledge* (Polanyi, 1958), in a different form possibly in Lonergan's *Insight* (Lonergan, 1957), and with explicit reference to hermeneutics, or the theory of interpretation, by Gadamer in his well-known controversy with Habermas (Gadamer, 1976). By now, with the spreading acceptance of hermeneutics as a fundamental perspective for epistemology, it is becoming clear to a numerous, if rather oddly assorted, collection of people that science as such, and certainly the study of complex systems, necessarily entails interpretation. Gunther Stent goes so far as to argue in this book that science is hermeneutical because the central nervous system itself is hermeneutical. I prefer to start from perception rather than from neurology, but the point is the same. The perceiver, taking note of things or events in his or her surroundings, must already be equipped to pick up the meanings of such things and events. The perceiver's reading of those meanings, of the text of nature, educates him or her—literally leads him or her forth—to a further, enriched reading, which in turn leads to a still more enriched reading, and so on. The hermeneutical circle within which we find ourselves, and through which we are in contact with the world, both enables us to gain, criticize, refine, and elaborate knowledge about things and events within the world and limits the forms of such knowledge through the contingent givens of our habits of thought as well as through the contingent forms that nature itself has acquired. That is the ultimate co-evolution within which all our science necessarily develops—or declines. Pure objectivity is thus an illusion. But that is not to say that science is subjective either, a matter of fashion, convention, or sheer arbitrary choice. Objectivity, like judicial impartiality, is a standard that is genuinely authorized within certain societies. It is disinterestedness, which is itself an interest, the target of care. Like the stamina to run a marathon, or the persistence to get through medical school, or the ingenuity needed to devise experiments, it has to be cultivated and sustained by an arduous training program.

Science as a Form of Life

From the primacy of history and the pervasiveness of interpretation follows my assertion that sciences should be viewed, not as logical systems but as "forms of life," in Wittgenstein's phrase, or, to follow A. C. MacIntyre, as "practices." By a practice, MacIntyre means,

. . . any coherent and complex form of socially established cooperative human activity through which goods internal to that form of activity are realized in the course of trying to achieve those standards of excellence which are appropriate to, and partially definitive of, that form of activity, with the result that human powers to achieve excellence, and human conceptions of the ends and goods involved, are systematically extended. . . . Tic-tac-toe is not an example of a practice in this sense, nor is throwing a football with skill; but the game of football is, and so is chess. Bricklaying is not a practice; architecture is. Planting turnips is not a practice; farming is. So are the enquiries of physics, chemistry and biology, and so is the work of the historian, and so are painting and music (MacIntyre, 1981, p. 175).

If we accept this definition, and the identification of sciences as practices, what does this mean for our philosophical interpretation of science? It means, as I have been emphasizing all along, that science is something people do. It is not, as such, a logical structure, but a human pursuit, a complex, intrinsically organized as well as extrinsically authorized network of activities. Thus science, like any practice, has its own inbuilt standards or values: scientific values.

There are various lists of such scientific values in the literature. Polanyi mentioned accuracy, scope, and intrinsic interest (Polanyi, 1958). Thus a highly mathematicized and far-ranging discipline like physics gains in accuracy what it loses in intrinsic interest. There is a lot that matters that it cannot tell us about at all. On the other hand, evolutionary theory has great intrinsic interest, because it tells us about the origin of living things, including ourselves; and in a way it also has wide scope, though not as wide as that of the laws of physics. But it is still plagued by ambiguities, as it has always been. The mathematics it uses is not terribly sophisticated, and by itself the mathematics does not tell us anything. Indeed, evolutionists, to get anywhere, have to be good biologists, and that is a very complex, subtle, yet imprecise sort of critter to be.

Richard Levins gives a similar list of desiderata for biological models: realism, accuracy, and scope. He explains that it is nearly impossible to achieve all three at once in any one explanatory model (Levins,

1966). Thomas Kuhn, in one of the essays in *The Essential Tension*, lists accuracy, consistency, scope, simplicity, and fruitfulness as scientific values (Kuhn, 1977, p. 322). Implicit elsewhere in his discussions, and in other accounts also (as, again, in Polanyi's), is what one might call novelty or possibly originality, although that is perhaps too grand a word. I mean just that no report of a piece of research gets accepted by the scientific community if it fails to present something new. This follows, in fact, from my point, that science is progressive problem-solving. To have command over past, even present, scientific knowledge, is to be a historian of science, not a scientist. To be a scientist, one needs to turn that knowledge to account in the search for more knowledge.

So, again in opposition to the older philosophy of science, which contrasted facts with values, objective with subjective, the new philosophical account insists on the value-laden character of science. Science, like any practice, includes intrinsic values, the adherence to which makes it the practice it is. What distinguishes science as a practice, or better, as a family of practices, from baseball, farming, or architecture, is that it seeks truth. This goal it shares with history, although their methods as well as the nature of the truth they seek are different. It may be argued that the arts too seek truth. Certainly, we understand our experience more richly through the arts. That is so. But the goal of truth, and especially the truth about some natural phenomenon, as distinct from the nature of human experience as such, is the focal and explicit aim of science in a way that does not hold for the arts. Art teaches, but is not didactic. Science is intended to teach both the scientist himself or herself and all those willing and able to learn what, in some respect and within some accepted theoretical terms, is the case. Helmuth Plessner once remarked that human beings construct different sorts of artifacts: languages, religious rituals, buildings, works of fine art, theories. But these have different targets, different reasons for existing. A work of art, however much it tells us about nature or our nature, is not designed for the conveyance of information. Breughel's *Fall of Icarus*, for example, shows us, if you like, the insignificance of human aspirations compared with the realities of nature or ordinary life. But the painting is not aimed solely at that message. In every aspect of its composition and in its whole being it has, as Jakobson has remarked of poetic discourse, "its own weight and its own value." Just the contrary emphasis holds of scientific discourse. However beautiful or elegant in their articulation they may be, scientific theories, explanations, or experiments are aimed at telling those who can read them how things are. In a famous statement the physicist Dirac declared, in effect, that if he had found a beautiful theory no one could validate adequately through experiment, he would look for a more beautiful theory, confident that it would prove true (Dirac, 1963). Norman Campbell reflected the same attitude when he described a scientist as a person who believes that nature will conform to his or her intellectual desires (Campbell, 1952, p. 89). But again, in this context, it is the conformity, not the beauty or the desire, that provides the payoff. Like everyday language, scientific theories, scientific explanations, scientific laws are (relatively) transparent, aiming at some real phenomenon in nature itself. And it is just that theoretical outward direction that characterizes the sciences as practices, unlike, say, the practice of painting.

The Social Nature of Science

If sciences are practices, it follows that they are social structures in at least three senses. For one thing, sciences exist as social enterprises. They demand what Polanyi called "conviviality." That is not to say that science as such is a wholly cooperative venture in which objectivity reigns, and all together meekly bow before the facts, or all together venture on a hazardous experiment in the spirit of a storybook surgeon with his loyal team. Of course the life of science, like any form of life, is characterized by competition, fear, and greed. Indeed, the three passions listed by Hobbes (after Thucydides) as characterizing life in society—fear, greed, and (vain)glory—are as typical of the scientific as of any other human activity. (Fear may not be as obvious as the others. But fear of one's superiors, or even of one's peers, is surely a factor in shaping what one may or may not say, do, or publish, from one's dissertation to one's hope of a Nobel Prize. Fear is always the reverse side of hope.) And, of course, the life of science not only includes antisocial social motivations, but is also, in one sense, nonsocial, individual, in that only A knows what A knows, B what B knows, and so on. Knowledge is open to all, but each one acquires, in his or her own person, only as much as he or she does, in fact, acquire. Even the closest cooperation, like the closest love or friendship, is essentially the union of two or more with one another, not quite e pluribus unum. All the same, as the sciences have grown in complexity, and even in bureaucracy, what Holton calls "public science" has come to replace almost altogether its predecessor or alternative, the science of the beginner or amateur (Holton, 1973, pp. 19–24). Public science here means science supported by institutional structures, whether its funding comes from government or from the private sector. In either case science is public and communal, as distinct, say, from the idiosyncratic tinkering of a British gentleman microscopist in the eighteenth century. Earlier still, even a relative solitary like Leeuwenhoek, with his "little animals," depended on others before him and in his own time who had been, and were, developing the art of lens-making. But by now the practices making up the complicated network of interlocking activities that sum up to science are plainly supported by a very subtle and delicate web of interactions constitutive of their very existence as components of the scientific world.

This conception of science as social in its very nature contrasts once more with the received view, which saw science not only as utterly impartial and in that sense impersonal but also as wholly outside any contact with personal, let alone interpersonal, aspiration or achievement. The scientist in the laboratory was supposed to be a pure automaton, a carrier whose individual life was as unimportant to the structure of scientific knowledge as Typhoid Mary was, in herself, to the symptomatology of the illness she spread. And if the scientist is an irrelevant nonperson, a fortiori his own or her interaction with other such nonpersons would be doubly irrelevant to the nature of science as such. Science, it was held, is a logical structure; society, on the other hand, is a collection of people—what have they to do with proof, deduction even observation? Clearly, the alternative I have been adumbrating demands an end to such intellectual apartheid. Every practice by its very nature constitutes a society with its own internal values; its structure is a social structure, from first to last.

A second aspect of the point I am making here—that is, of the sociality of science—is that there is a particular form of sociality entailed in science. For sciences, like other practices, have to be taught and learned. They are traditions, including, paradoxically, a tradition of criticism of the tradition. They require apprenticeship, the assimilation of the young scientist-to-be to the standards, techniques, and vocabulary of his or her discipline. Michael Polanyi used to say that if you want to find out what the fundamental principles of a science are, you should spend five years in the laboratory of a master of that discipline. The acquisition of mathematical, usually experimental, sometimes field observational techniques, the internalization of a given technical language, the feel for the right question, the sense of what to do next—all that belongs to the repertoire the novice has to acquire, so that he or she, qua scientific learner, becomes, in part, a different person. This is indeed the interpersonal aspect essential to the hermeneutical circle that lies at the foundation of any practice, including the sciences. The practice of learning from masters, and so of entering into a new intellectual and technical world, is something the older philosophy of science had necessarily to ignore. On that model, again, science was simply apart from society. Such a relation as that of teacher to learner was irrelevant to its simon-pure logical ideal.

A third and final aspect of the social nature of science is that the sciences not only form an interlocking network of communities, in which the master-disciple relation plays a particular role, but that the sciences as a whole are also part of society. They need the authorization and the respect of society as a whole in order to come into, and to continue in, existence. According to the former orthodoxy, science was held to be either apart from society or superior to it. If the latter, the inference was: if we would only be scientific about our problems—all our problems, personal, political, economic or what you will—we could solve them all. Leibniz entertained the hope that, given the universal notation he intended to devise, people disagreeing about any issue would sit down with a slate and say, "let us calculate." Computers are more elaborate than slates, but the idea is the same. In fact, however, it is the contrary that is the case. If scientific institutions and scientific knowledge are to be permitted to develop, society has to clear a space for the pure pursuit of knowledge, to recognize its value not only for its practical spinoff, but in itself. This is a question not just of grants-manship, of money-getting, although that too, alas, is a necessary con-dition, but of respect. The sciences, like other practices, have their own internal values, but they also need, for their sustenance, to be valued, and therefore authorized, by the society within which they find their place. If salvation, or money-making, or entertainment becomes the exclusive goal of a society, science must wither away.

Nor, it should be noted in passing, can science, by its very nature as a practice, supervise all the tasks of a society, as Leibniz believed, and as some scientists still believe. Because each practice has its own aims and standards, none is or can be comprehensive in the sense that it is competent to deal with the objects of each and every human need. In particular, science is subject to society and not directive of it because its interests are limited and abstract. The concerns of human beings cannot, on principle, be formulated, let alone resolved, through the relatively impersonal and nonpersonal standards constitutive of the natural sciences. Even though all disciplines are contained within their own hermeneutical circles, the interpretive frameworks of disciplines differ from one another. And precisely because their object as well as their subject is human, the evaluative component of the human sciences is deeper and richer than is the case for the sciences of nature. (I cannot go into this delicate distinction here; it is dealt with in some detail in Polanyi's *Personal Knowledge* and its rationale is clearly set forth in Taylor's classic paper, both already referred to). In other words, although the hermeneutical circle characterizes all the sciences, it characterizes the so-called human sciences in a peculiarly rich and irreducible way. Nor, indeed, can even the human sciences, qua science, embrace the most immediate questions of personal or political life. But that is yet another story.

Scientific Pluralism

The pluralism of practices leads me to my final point. An aspect of the older orthodoxy, defended by many of its adherents, concerned a program for the "unity of science." If there is *the* scientific method, it was thought, there should also be, when science eventually matures sufficiently, *the* one body of science itself, with its one unified set of axioms, from which ideally all the empirical content of science can be deduced. If sciences are practices, on the contrary, not only is this one family of practices distinguished from others, but it comprises an interlocking network of disciplines, each impinging on, but different from, its neighbors. If one becomes a biochemist, one cannot at the same time and by the same training become an astronomer. A neurophysiologist has no competence in paleobotany. Nor is this a case of stultifying specialization. Expertise is essential to every practice. At the same time, neighboring disciplines do impinge on one another. As Darden and Maull have argued in their authoritative essay, interfield theories form an important ingredient in many areas of research (Darden and Maull, 1977). In these cases, neither field is being reduced to the other, let alone to some universal bottom-line science. Obviously, for example, geometrical and physical constraints operate in the development and structure of organs or organ-systems. Thus, some geometrically and physically construed structures are, in certain situations, biologically more advantageous than others (Beatty, 1980). But that does not mean that anatomy, physiology, or evolutionary theory are thereby reduced to physics or geometry. The techniques of the sciences are many and subtly various. Pantin distinguished biology and geology as unrestricted sciences from the exact and more restricted disciplines, since the former draw more strikingly for their information and techniques on a number of other branches of science (Pantin, 1968). Yet this distinction too is probably one of degree. The point is that, on principle, there is not science as such; there are disciplines, mutually interrelated and interacting, that develop as distinctive scientific practices.

One warning, finally, in conclusion. Sciences are practices essentially aimed at knowledge. Disappointed in the hope of eliciting or elucidating a pure hypothetico-deductive or deductive-nomological account of science from the givens that face them, some writers have recently adopted what they call a "strong theory" in the so-called sociology of knowledge (Bloor, 1976). There is no difference, they say, between a true and a false discovery. Nonsense! A false discovery is no discovery. And a genuine discovery is so just because there is good reason to hope the statements announcing and confirming it are true. It may turn out to need modification, even radical modification. But if, like Stahl's discovery of phlogiston, it turns out to be totally mistaken, then it was an alleged discovery and so no discovery at all, but, as we are now justified in asserting, an error. As science is infected with human frailty, so the truth it aims at, as its primary and constitutive value, is infected with error. As the search for knowledge that constitutes science may be corrupted by ambition and dishonesty, so the knowledge science aims at—and, we are justified in believing, often succeeds in attaining—may fail of fulfillment. Like any human activity, science is fallible. That does not mean, however, that the passion for knowledge, an intellectual passion, is not its fundamental support, or that that passion is not sometimes fulfilled (Polanyi, 1958, pp. 132–202). To legislate the claim to know out of existence because it is corruptible, or because it may succeed or fail, is to fail to grasp either the character of practices in general and their role in human history, or the character of scientific practice in particular. True, science is something people do or it is nothing at all. But, more than that, science is something people do in order to try to find out the truth about some natural process—or it is not science.

I have gone on at some length about this, since the collapse of the older view has led to such an extreme swing toward a noncognitive account of science that it is important to prevent our falling into that trap. The same holds, of course, for the contrast mentioned earlier between the conception of science as progressive problem-solving and the extreme incommensurability notion that tended to succeed the collapse of the once received ahistorical view. The option I am proposing, it cannot be too heavily stressed, does not constitute an abandonment of the reality of science as a collection of cognitive achievements or of the authority of science in its appropriate sphere, but an effort to see science as real and as authoritative because it is an important collection of human activities pursued out of motives we consider honorable and with results we respect and admire. Above all, it is a reasonable and rational pursuit, of which we can give a reasonable, and indeed a rational, account, even though—or just because—it is historical and social in its existence and its nature. It is also, and just as fundamentally, as we noticed first of all, rooted in the effort to make out through the use of one's perceptual systems the real structures of the world in which, bodily, we have our being. Such efforts can and should be studied not as the blind lashing-out of noncognitive drives, or, as some would have it, efforts at self-perpetuation by our genes, but as reasonable efforts to discover and, when they succeed, to know how something in the real world really happens, really is. It is comprehensive realism the new philosophy of science is seeking to articulate. All the other theses I have listed should be subsumed under that fundamental claim. Why such a comprehensive realism should be especially appropriate for biological-philosophical questions, I trust that I have indicated at the start.

References

Beatty, J., 1980. Optimal design models and the strategy of model building in evolutionary biology. *Phil. Sci.* 47:532–561.

Bloor, D., 1976. *Knowledge and Social Imagery*. London: Routledge and Kegan Paul.

Brown, H., 1977. *Perception, Theory and Commitment*. Chicago: University of Chicago Press.

Campbell, N., 1952. *What Is Science?* New York: Dover.

Compton, J., J., 1983. Natural Science and being-in-the-world. Paper read at Pacific Division, Amer. Phil. Assoc., March, 1983.

Darden, L., 1983. Reasoning in theory construction: analogies, interfield connections and laws of organization. *Abstrs. 7th Int. Cong. Logic Meth. and Phil. Sci.* 4:288–291.

Darden, L., and N. Maull, 1977. Interfield theories. *Phil. Sci.* 44:43–64.

Dirac, P. A. M., 1963. The evolution of the physicist's picture of nature. *Sci. Am.* 208(5):45–63.

Gadamer, H. G., 1976. *Philosophical Hermeneutics*. Berkeley and Los Angeles: University of California Press.

Gibson, E. J., 1984. The concept of affordances in development. The renascence of functionalism. *Minn. Symp. on Child Devel.*

Gibson, J. J., 1979a. The Ecological Approach to *Visual Perception*. Boston: Houghton-Mifflin.

Gibson, J. J., 1979b. The gap in the ecological approach. Dittoed note in the Cornell Archives (quoted by permission of E. J. Gibson).

Gould, S. J., 1983. The hardening of the synthesis. In *Dimensions of Darwinism*, M. Grene, ed. Cambridge and New York: Cambridge University Press, pp. 346–360.

Gregory, R. L., 1966. *Eye and Brain*. New York: McGraw-Hill.

Grene, M., 1974. People and Other Animals. In *The Understanding of Nature*, M. Grene, ed. Dordrecht: Reidel, pp. 346–360.

Grene, M., 1983. Empiricism and the philosophy of science, or *n* dogmas of empiricism. In *Epistemology, Methodology and the Social Sciences*, R. S. Cohen and M. W. Wartofsky, eds. pp. 89–106.

Heelan, P., 1983. *space Perception and the Philosophy of Science*. Berkeley and Los Angeles: University of California Press.

Heidegger, M., 1927. *Sein und Zeit*. Halle: Niemeyer.

Holton, G., 1973. *Thematic Origins of Scientific Thought, Kepler to Einstein*. Cambridge, Massachusetts: Harvard University Press.

Kuhn, T., 1977. *The Essential Tension*. Chicago: University of Chicago Press.

Levins, R., 1966. The strategy of model building population biology. *Amer. Scient.* 54:421–431.

Lonergan, B. J. F., 1957. *Insight: A Study of Human Understanding*. New York: Harper & Row.

MacIntyre, A. C., 1981. *After Virtue*. Notre Dame, Indiana: Notre Dame University Press.

Maull, N., 1977. Unifying science without reduction. *Stud. Hist. Phil. Sci.* 8:143–162.

Mayr, E., 1982. *The Growth of Biological Thought*. Cambridge, Massachusetts: Harvard University Press.

Merleau-Ponty, M., 1942. *La Structure du Comportement*. Paris: P.U.F.

Merleau-Ponty, M., 1945. *La Phenoménologie de la Perception*. Paris: Gallimard.

Pantin, C. F. A., 1968. *The Relations Between the Sciences*. Cambridge: Cambridge University Press.

Polanyi, M., 1958. *Personal Knowledge*. London: Routledge and Kegan Paul; Chicago: University of Chicago Press.

Provine, W. B., 1983. The development of Wright's theory of evolution: systematics, adaptation and drift. In *Dimensions of Darwinism*, M. Grene, ed. Cambridge and New York: Cambridge University Press, pp. 43–70.

Riegl, A., 1901. *Die Spätromische Kunstindustrie*. Vienna: K. K. Hof. and Staatsdruckerei.

Taylor, C., 1971. Interpretation and the sciences of man. *Rev. Met.* 25:3–51.

Zerner, H., 1976. Alois Riegl: art, value and historicism. *Daedalus* (Winter 1976): 177–188.

On Conceptual Change in Biology:
The Case of the Gene

Richard M. Burian

The current situation in philosophy of science generally, and in philosophy of biology in particular, is most unsatisfactory. There are at least three general problems that many philosophers thought themselves near to solving twenty years ago, only to find that the anticipated solutions have come unglued. These are (1) the problem of characterizing and understanding the dynamics of conceptual change in science; (2) the problem of understanding the interrelationships among theories (including particularly the reduction of one theory to another); and (3) the problem of scientific realism (i.e., the problem of how seriously to take the claims of theoretical science or, at least, of some theoretical scientists, to be describing the world—literally—in terms of such theoretical entities as genes and protons, DNA molecules, and quarks). This general situation has significant effects on the philosophical study of particular sciences. In philosophy of biology, for example, although one finds a large number of elegant studies of particular topics, the sad fact is that there is no generally satisfactory large-scale synthesis in sight. We have no agreed-on foundation, no generally acceptable starting point from which to delimit and resolve the full range of theoretical problems of interest to scientists and philosophers regarding biology.

This chapter provides a preliminary report on a new approach to conceptual change, together with a sketch of its application to important biological subject matter. The approach offers some promise of providing a satisfactory framework, compatible with scientific realism, for detailed studies of particular scientific developments.

Before I sketch in some of the relevant philosophical background, it will be useful to indicate how various concepts of the gene will enter the discussion. It is now 84 years since Mendel's work was rediscovered. Within fifteen years of that event, say with the publication of *The*

Mechanism of Mendelian Heredity by T. H. Morgan and his coworkers in 1915, the main elements of the classical theory of the gene were fairly well established. A nearly constant series of improvements and refinements in that theory, grounded in good part in laborious but fascinating experimental work, resulted in considerable revision of that theory. Indeed, the accumulated changes run so deep that some have characterized this historical process as one in which Mendelian genetics was replaced by a series of improved successors which can be grouped under the label *transmission genetics*. (For one example, cf. Hull, 1974, chap. 1.) This extended process, both in its theoretical and its empirical aspects, helped prepare the way for what is usually considered to be (to use the vogue label) a scientific revolution brought on by the advent of molecular genetics. For present purposes, we may mark that advent by the publication of the justly famous solution of the principal structure of DNA in Watson and Crick (1953).

It will not be possible to review the relevant history in this chapter—the task is simply too large.[1] Instead, I will single out a couple of moments from the history of genetics to illustrate the character of conceptual change in that discipline and to support the claim that the conceptual changes examined fit well with the larger philosophical views put forward in this chapter. I will show that the approach which biologists took to the conceptual changes in question had a significant effect on their practice, a result which suggests that the proper handling of conceptual change ought to be of real concern to working scientists. Finally, I will suggest that a full-scale study of the concept of the gene is a singularly appropriate vehicle for working out a general account of conceptual change in science.

Continuity and Discontinuity in Scientific Theories

A very crude description of the present impasse in the theory of conceptual change will suffice for present purposes.[2] In spite of an immense variety of refinements, there are two main views to be considered. One of these, associated with the names of Paul Feyerabend[3] and (perhaps mistakenly) Thomas Kuhn[4], might be labeled the discontinuity view. It claims that there are, at least occasionally, genuine conceptual revolutions in science. Revolutions are to be understood in a radical way; when they occur, those scientists who are separated by a revolution in a given field end up working with concepts and theories that are mutually incommensurable. This is to say that the pre- and post-revolutionary theories and concepts do not share a common denominator, and so are not interdefinable in any useful way. Discontinuity theorists usually add that the recording of scientific observations requires some sort of conceptual apparatus and that the concepts involved in gathering and reporting scientific observations are theory-laden in some way or other. With these additions, they argue, it follows that observations worked up to support or test one of the theories in question need not (and in difficult cases will not) serve in any direct way to support, test, or undermine the competing theory. Otherwise, the relevant observational concepts could be used to provide at least partial interdefinitions of the concepts drawn from the competing theories. But those theories were supposed, by hypothesis, to be incommensurable.

This result is counterintuitive. At first glance, at least, it seems simply incorrect to claim that evidence gathered under Newtonian or Mendelian auspices must automatically be thrown out or recast if it is to have a bearing on quantum mechanics or on molecular genetics, respectively. Not surprisingly, such claims have been highly controversial, with some thinkers supporting them and others arguing that they reduce discontinuity views to absurdity. Indeed, the best-known theorist of scientific revolutions, Thomas Kuhn, has backpedaled a long way from this sort of reading of his work in an attempt to avoid the "absurdities" that result from extreme interpretations of incommensurability. (Cf. Kuhn, 1983b.) But such extreme interpretations are not easily avoided. The difficulty is that people (including Kuhn, who, as will be shown in n. 6, even now cannot escape the central difficulty) arrived at discontinuity theories honestly, by means of a series of persuasive arguments which have not been adequately answered.

Discontinuity theories arise from a set of seemingly commonsensical commitments regarding the nature of scientific language, concepts, and theories. Among these is the notion that a scientific theory is adequately represented as an interconnected body of statements about some domain of phenomena, and that this body of statements typically employs characteristic theoretical concepts. Theoretical concepts, in turn, cannot be properly understood in abstraction from the relevant theory. While it may, arguably, be true that one need not employ the whole of a theory to understand or delimit the relevant concepts, one does presuppose at least some central core of the theory in question when utilizing those concepts. Thus, to measure the mass of a body, one must presuppose at least Newton's second law ($F = ma$), and probably the third law as well. And for 'mass' to refer, i.e., for bodies to genuinely have mass, bodies must in fact behave in accordance with Newton's laws; in short, the fundamental laws of Newtonian mechanics must be true if their theoretical terms are to refer to objects or properties in the world.[5] Similarly, to determine whether an organism is heterozygous for a recessive Mendelian gene, say a gene for albinism, a classical geneticist must presuppose various theoretical principles. One of these principles claims, for example, that, other things being equal, an organism that has a recessive gene in double dose will, in the right circumstances, exhibit the trait (in our example, albinism) in terms of which the gene is identified. Obviously, on any reasonable account of the structure of classical genetics, further principles concerning the transmission of genes from parents to offspring must also be presupposed in setting forth the workings of heredity. And organisms will possess genes only if the relevant Mendelian laws are true (or, perhaps, only if some fundamental subset of those laws is true).

These considerations show that a holistic account of theoretical language has important consequences. According to a holistic account, the proper employment of a theoretical concept requires one to mobilize some fairly large entity—an entire theory or paradigm or conceptual scheme or conceptual framework or theoretical language or world view, to borrow some of the terms current in the literature. Such holism brings radical incommensurability with it. What has led (or sometimes forced) people to swallow discontinuity theories has been their acceptance of some form of holism about theoretical concepts. And they have been forced to this holism, in turn, by recourse to wholly inadequate semantic theories for theoretical languages.

The reason for claiming that Kuhn, like the rest of us, has not solved this tangle of problems is that he has not shown how to integrate his account of conceptual change into a nonholistic theory of theoretical language. What is required is some way to spring concepts like 'mass' and 'gene' sufficiently loose from the fundamental principles of the theories with which they are allied that it becomes apparent how one could seriously use those terms as a scientist without presupposing the truth of the corresponding principles. Only then will we be able to make sense of debates between theorists over which of their theories, built on incompatible principles, is a (or the) correct theory of mass or of the gene. It is not enough to wish to reject extreme versions of incommensurability; until philosophers show how to get around holistic theories of the language of theoretical science, they will be vulnerable to arguments forcing them into acceptance of radical incommensurability.[6]

For the sake of completeness, I shall deal briefly with the second class of theories of conceptual change in science, namely continuity theories. At the moment, most extant continuity theories are discredited, for they are committed to the idea that some core of scientific concepts (for example, so-called observational concepts) provides a permanent (though perhaps expandable) base for the spinning out of new theoretical concepts and theories. One reason for the disrepute into which continuity theories have fallen is their close connection to traditional philosophical accounts of the reduction of one theory to another. These accounts, alas, are a total failure. They claim that when a theory (say Mendelian genetics) is reduced to (or by) another (say molecular genetics), one can deduce the central claims of the theory which is reduced from those of the new, more fundamental theory.[7] The deduction uses the principles of the fundamental theory together with suitable definitions and statements connecting the concepts of the two theories, and descriptions of the circumstances in which the reduced theory obtains. Were this account of reduction correct, the concepts of the reduced theory would be, in effect, definable within the more fundamental theory and the claims of the reduced theory would be a subclass of the claims of the fundamental theory. In fact, as is now generally recognized, reduction of this sort virtually never occurs in science. One cannot even deduce Kepler's laws for our solar system from Newtonian mechanics. Given that there is more than one planet, what one deduces is the claim that the orbital formulae obtained from Kepler's laws are approximately correct though literally false; the calculable perturbations of Mars from its supposed elliptical orbit, for example, were observable within the limits of accuracy with which Kepler was working. Worse yet, when one comes to cases like statistical versus phenomenological thermodynamics or, as we shall see, Mendelian versus molecular genetics, discontinuity theorists have put forth quite convincing arguments to show that the concepts of the theory to be reduced simply cannot be reproduced within the successor theory. (Hull, 1974, chap. 1.) This claim, as I shall argue, seems to be entirely in accord with the facts.

Over and above the connection between continuity theories and discredited theories of reduction, there is fairly broad consensus that they are not true to the facts. There simply is no timeless and secure set of observational concepts, let alone nonobservational concepts, strong enough to be employed as the base for building up the concepts of theoretical science.

How do all these abstract considerations bear on the workings of real science? What on earth is to be accomplished by dismissing both continuity and discontinuity theories of conceptual change? For those who wish to surmount difficulties, not just to find them, the next steps may provide some relief. I shall argue that the standard theories of conceptual change are vitiated by a mistaken presupposition that prevents their having real contact with live science. An examination of the concept—or rather the concepts—of the genes provides an ideal vehicle for bringing about a new start on a much more satisfactory footing.

The Gene: Development of a Concept

It is time to make a first pass at the various concepts of the gene. The question which we will pursue is this: to what, if anything, does a scientist refer when he uses the term 'gene' or one of its cognates? For present purposes this question will provide enough of a guiding thread to lead us through what will someday have to be turned into an immensely complex discussion.

Honesty compels me to remind the reader how complex a full treatment of the history of the gene will have to be. A single sentence from Carlson (1966) makes the point:

The gene has been considered to be an undefined unit, a unit-character, a unit factor, a factor, an abstract point on a recombination map, a three-dimensional segment of an anaphase chromosome, a linear segment of an interphase chromosome, a sac of genomeres, a series of linear sub-genes, a spherical unit defined by target theory, a dynamic functional quantity of one specific unit, a pseudoallele, a specific chromosome segment subject to position effect, a rearrangement within a continuous chromosome molecule, a cistron within which fine structure can be demonstrated, and a linear segment of nucleic acid specifying a structural or regulatory product. (p. 259)

To cut through all this complexity, I shall make four fundamental points by use of one rather simple example. The first point is that it is possible for scientists to exercise strong controls to ensure that they are referring to the same entity or entities in spite of very large differences in viewpoint, terminology, concepts, and theoretical commitments. The episode in terms of which I will make this point concerns William Bateson, one of the founding fathers of genetics—the one, in fact, who coined the very term *genetics*.[8] Bateson never was fully persuaded that genes could be localized on chromosomes or that they could be mere molecules or material particles. (Because of the dynamics required for them to achieve their effects, he thought that they would have to be stable harmonic resonances or something of the sort.) He usually employed the term 'factor' rather than the later coinage 'gene', and he used 'character' or 'unit character' for traits which he counted as the effects of the presence, absence, or, rarely, alteration of a single factor. He preferred to talk of 'discontinuous variation' rather than 'mutation' and he distinguished sharply between such discontinuous variations (which he thought were the real stuff of evolution) and the small continuous variations which he took to be the target of Darwinian natural selection.

Bateson (1916) is a review of the Morgan group's definitive book, *The Mechanism of Mendelian Heredity*, published by *Science*.[9] Now Bateson and Punnett had discovered linkage between genes (which they called 'gametic coupling') in 1906. Ironically, linkage (which, by the way, requires violation of Mendel's law of independent assortment) served as one of the cornerstones for the Morgan group's arguments that chromosomes are the bearers of the hereditary material and that the relative locations of the factors or genes may be pinpointed quite precisely by means of maps based on the degree of linkage between them. The point about the Bateson review is very simple: in spite of all the differences between his views and those of Morgan and his colleagues, there simply is no difficulty about the reference of the relevant terms. Without explicit consideration, Bateson considers it to be established that certain traits (e.g., certain eye color traits in *Drosophila* which had been shown in experimentally controlled matings to be transmitted to offspring in well-established particulate patterns) are indeed the consequence of the presence (or perhaps, sometimes, the absence) of a particular factor. The main disagreements in question concern not whether there are factors or genes acting in such cases, but rather what kinds of things these factors are and how strongly the available evidence supports the view that they are material particles (or some such) located at the relative positions on the chromosomes worked out by the Morgan group's mapping techniques. In the much later operationalist terminology of L. J. Stadler (1954), Bateson has retreated to an operational definition of the gene which he shares with Morgan's group; his disagreement with them concerns the best theoretical account to offer of the behavior of the operationally defined gene. In Stadler's terminology, this disagreement concerns the hypothetical gene. (I will return to Stadler's distinctions briefly at the end of this chapter.)

The second point to be drawn from this episode is that the procedures involved cannot be confined to Bateson's and Morgan's groups; they are already the firmly established property of a larger community. It is, of course, true that the procedures by which factors (or genes) are identified and individuated can be refined and improved in ways which may shift the reference of particular terms; such refinements can change altogether the set of entities to which a community refers by use of such terms as 'gene' and 'factor'. But the terms are, in a certain sense, community property. There are experts who are in a position to adjudicate questions regarding whether or not Morgan and his colleagues met the conditions required for them to be referring, at least prima facie, to particular genes and to determine whether or not any objections to their claim so to have referred are cogent or not. Bateson, himself one of those experts, knew that the Morgan group had unassailably demonstrated that they were dealing with single genes by the standards then available.

This point about the social character of the referential use of scientific and prescientific terms is rather stronger than it looks. As a concrete example which, in spirit, goes back to the work of Hilary Putnam (1975, pp. 223–229ff.) I am able to use the terms 'mole' and 'mole rat' to refer to distinct mammals in spite of the fact that if someone were to bring a few of each of them to me I could not tell which were the moles and which the mole rats. The reason that my lamentable ignorance does not prevent me from using these terms to refer correctly to distinct animals is that my usage is interlocked quite deliberately with that of a larger community, and particularly with that of experts, of whom I am prepared to defer, who can tell moles from mole rats. The matter is, of course, not so simple when it comes to questions on the forefront of knowledge; all of the experts may be (and have on certain issues been) wrong in their central beliefs about genes. Nonetheless, as the very example of the Bateson review demonstrates, when the social controls work correctly and the world cooperates (neither of which can be counted on), the experts' procedures allow them to secure reference for theoretical terms like 'factor' and 'gene' even in the face of considerable disagreement, even in the face of pervasive and fundamental error in the foundations of their theories.

The third point suggested by consideration of the Bateson review is that terms like those under discussion are used by the community in a way which does not require that their reference be fully specified. To put the point differently, the account we give of the reference of terms like 'gene' ought to allow radically different descriptions of the things which genes might turn out to be. We often secure reference for theoretical terms and concepts while using deeply mistaken theories. It is clear, I hope, that if the world were appropriately different and if the genes in question in the Bateson review were in fact stable harmonic resonances rather than segments of chromosomes, nothing would need to be changed in what Bateson or Morgan group wrote, but the reference of some of their terms would have been different. To use a term from Kitcher (1978, 1982), we must consider not only the reference, but the reference potential[10] of a term to understand how it is used. One can argue that, as of 1916, the terms used to refer to, say, the vermillion eye color gene of *Drosophila melanogaster* included stable harmonic resonances in their reference potential. Not long afterward, thanks to the Morgan school's success, that reference potential shifted; some genes, at least, had to be localized on chromosomes. Accounting for such changes is an important part of a satisfactory history of the concept of the gene.

The fourth point is more or less a corollary of the third. It is sometimes possible to resolve disputes regarding the reference of theoretical terms in the light of subsequent investigation. No matter what Bateson thought he was referring to, when he spoke of the *Drosophila* genes, he was in fact referring to segments of *Drosophila* chromosomes. Sometimes, at least, the world is well enough behaved to allow a clean resolution of such disputes. At the same time, to gain a proper appreciation of Bateson's writings, we must not let the truth of the matter run away with us. Bateson's beliefs inevitably affected his terminology; in this he is typical of all working scientists. Sometimes he used the term 'factor' in the attempt to refer to the entities which his own undeveloped theory, if correct, would have described. In such instances (since there are no such things) the term does not, in fact, refer at all. At other times he used the term 'factor' to refer to those entities, whatever they might turn out to be, which he, the Morgan group, and many others had singled out by their experiments. The unwary reader who fails to distinguish these different uses of the same term will be unable to evaluate Bateson's claims correctly.

Theories of Reference and the Gene

Let us collect our results to date. From a philosophical point of view, they concern the theory of reference and the proper analysis of theoretical concepts. I shall take each of these in turn.

At the moment there are two fairly standard theories of reference to cope with. The traditional theory,[11] whose roots lie in the work of Frege, holds that kind terms like 'gene' refer, in the first instance, to any and all objects which fit the underlying description or descriptions with which the speaker is prepared to back up his or her use of the term. Those descriptions are said to spell out the sense of the term in question. According to the traditional theory, the reference of a kind term is fixed by its sense and by the world; a kind term refers to just those objects which fit the description or descriptions implicit in the sense of the term.[12]

In the context of the present investigation, this traditional theory can be seen to be allied with the holistic theories of theoretical language rejected above. Its natural application to the Bateson case employs Bateson's mistaken and unarticulated theory of factors to determine what the sense of 'factor' is and thus, in turn, to fix what he referred to in his primary uses of that term. The consequence is that the traditional theory claims that, precisely because Bateson's theory of factors (or genes) was fundamentally mistaken, Bateson did not refer to anything when he used whichever term for genes. On at least some occasions, in contrast, the Morgan group did refer to certain chromosome segments by use of such terms as 'gene' and 'factor'. But this contrast is wrong and wrongheaded. One small symptom of its erroneousness is that, taken seriously, it does not allow one to construct a plausible construal of Bateson's review.[13]

Although there are a number of moves that one can make to try to save the traditional theory in application to this case, this is not the occasion to explore them. I shall simply assert, dogmatically, that the traditional theory is wrong and ask its defenders to present their arguments to the contrary.

The alternative theory, various versions of which have been developed by Donnellan, Kripke, Putnam,[14] and others, is sometimes known as the causal theory of reference. (The label is a misnomer, but the issues involved are not of immediate importance to the matters at hand.) At first sight, the causal theory fares rather better in dealing with our case. It holds that what a kind term like 'gene' refers to depends on the earlier uses back to which present uses can be traced. 'Gene' refers to whatever natural kind (assuming that there is one) in fact entered into the relevant causal interactions when Mendel performed his experiments on garden peas and baptized the factors which determined the patterns of inheritance which those peas exhibited. Thus since, in the tradition we are exploring, the term 'gene' can be traced back to Mendel's uses of the German words *Charakter, Element, Faktor,* and *Merkmal* as means of describing the determinants of particulate inheritance, the causal theory says that both Bateson and Morgan were referring to the determinants of particulate inheritance, whatever their true nature.

Prima facie this theory of reference can handle the Bateson review. As usually developed, however, like the Fregean theory it is a closed rather than an open theory of reference.[15] By this I mean that the reference of such kind terms as 'gene' is treated as being completely fixed once those terms are properly established within the tradition. This closure, I believe, is yet another reflection of a holistic theory of theoretical language. Indeed, the fundamental presupposition shared by continuity and discontinuity theories of conceptual change is just this idea that the reference of theoretical concepts is closed. Closure of reference makes conceptual change into an all-or-nothing phenomenon.[16] When we turn to the history of genetics for the second time, we shall see that not only the reference potential but also the actual reference of the term 'gene' has changed in a controlled way during the development of the discipline of genetics.

But before turning again to cases, a few words are needed about reference potential. Following Kitcher (1978, 1982), though with important modifications,[17] I maintain that reference potential provides the central tool by means of which to analyze theoretical concepts. In the Fregean tradition, the concept mobilized by the use of a term is equated with the sense of the term, that is, with the underlying descriptions which speakers would employ to back up their use of the term. The alternative account that I am about to sketch is designed to reflect precisely those characteristics of the referential use of theoretical kind terms which give the Fregean tradition the greatest difficulty. Specifically, we need to take account of the ways in which the practice of scientists makes three things likely: first, that their fundamental theoretical terms will pick out some natural kind even when their theories about that kind are badly mistaken; second, that different scientists will be able to refer to the same natural kinds even when their theories about those kinds are in radical disagreement; and, third, that the precise referential use of such central theoretical terms as 'gene' can, in good cases, be brought into line with experimental results and with the theoretical commitments of the relevant community once that community achieves full consensus on those commitments.

Referring to Genes

The price that must be paid for theoretical discourse to have these characteristics is the referential openness of theoretical concepts and systematic ambiguity in the referential use of theoretical terms. Some will consider this price very high, but it has always been paid, for it is unavoidable. The history of genetics nicely illustrates the openness and systematic referential ambiguity of theoretical terms as well as some of the ways in which these are controlled. (The phenomena involved, though, are found in all of the sciences.)

Given the success of the Morgan group, the term 'gene' typically refers, in fact, to a segment of a chromosome which, when activated or deactivated, performs a certain function or has a characteristic effect. But how much of a chromosome? And what functions or effects? Much of the effort that went into mapping genes may be viewed as an attempt to answer the first question; much labor was expended on the determination of which part of which chromosome contained which genes. In the process, certain criteria were developed for telling one gene from another. According to one of these, if two mutations affecting the same phenotypic trait—say two eye color mutations—could be separated by recombination, then they belonged to separate genes; if they could not be so separated, then they belonged to the same gene. (That is, they were counted as alternative alleles at the same genetic locus.) This way of individuating genes was suggested by Sturtevant (1913a and 1913b), who suggested that two closely linked eye color mutations (called 'white' and 'eosin') that Morgan and Bridges had been unable to separate in an experiment using 150,000 flies (Morgan and Bridges, 1913) should be considered to be two alternative abnormal alleles at a single locus, a locus which had already been located in a specific position on the X chromosome. Now the more closely two genes are linked, the more difficult it is to separate them by recombination, and the larger the number of flies that must be used to execute the test. Thus it should be no surprise that such claims are sometimes wrong and that it was established many years later that, in this very case, one can separate the two genes in question if one performs a truly gigantic recombination experiment.[18] (Cf. Carlson, 1966, p. 64; Kitcher, 1982, p. 351.)

Consider the problem this creates when one asks what is referred to by subsequent uses of such terms as 'the gene for white eyes' or 'the eosin locus'. If one conforms to the usage established on the basis of Sturtevant's results, one refers to that portion of the chromosome which contains both the white and the eosin genes. But if one is working with the recombination criterion for theoretical purposes, one may refer, instead, to the smaller portion of the chromosome containing one, but not both of these genes. This is to say that two rather different segments of the chromosome belong to the reference potential of these phrases. Very often it makes no difference which portion of the chromosome one refers to. (They are, after all, virtually inseperable by ordinary techniques.) But occasionally it may matter whether one purpose or the other—conformity to established usage in order to accomplish coreference with other scientists or correct application of the criteria separating genes from one another—dominates one's usage. For a long time, the ambiguity was inescapably built into the mode of reference which was available in discussing these genes.

Indeed, at various stages in the history of genetics, it became a theoretical and practical necessity to distinguish between different gene concepts each of which picked out different segments of the chromosome or employed different criteria of identity for genes—remember Carlson's list! For example, in the 1950s Seymour Benzer pointed out that many geneticists had assumed that the smallest unit of mutation with a distinct functional effect coincided with the smallest unit of recombination— and he performed some elegant experiments which showed that this claim is false.[19] As a result, in some circumstances it became necessary to choose between the unit of function (which, for reasons which need not concern us, Benzer called the *cistron*), the unit of mutation (which he called the *muton*), and the unit of recombination (which he called the *recon*). This particular result showed that there had been hidden openness in the reference potential of the term (and the concept) 'gene' and that, in some arguments, though not in general, it was necessary to divide the reference of that term (concept) according to the separable modes of individuating genes.

The actual history is, of course, much richer than I have let on here, particularly when one pursues the story into the present, where one encounters split genes with separately movable subunits, transposable control elements, parasitic ("selfish") DNA, and so on. But enough has been said to show that there are at least four ways in which the reference of a particular use of the term 'gene', or one of its cognates, might be specified.[20] Which one of these is relevant will turn on the dominant intention of the scientist and the context of the discussion. One such intention is conformity to conventional usage. Taking Sturtevant's experimental result for granted, conformist usage would refer to the same segment of the X chromosome whether one spoke of the white or the eosin locus. Another, sometimes conflicting, intention is accuracy in the application of the extant criteria for identifying the relevant kinds or individuating the individuals of those kinds. When accuracy is the dominant intention, 'white' and 'eosin' refer to different segments of the chromosome. Thus Sturtevant's mistake expanded the reference potential of the term 'gene' by adding a compound chromosomal segment to the items potentially referred to by that term. In some, but only a few, contexts it provided terribly important to take the resultant long-unrecognized ambiguity of reference into account in order to understand the actual use of the relevant terms and to reconcile conflicts between competing descriptions of the outcomes of experiments. What is at stake here is the precise roles that one's theoretical presuppositions and accepted experimental results play in fixing the reference of one's terms. Although this discussion has not provided a general resolution of that difficult problem, it has given some indication of the proper apparatus to employ in carrying out case by case analyses.

The Benzer case illustrates a third way in which reference may be fixed: once an ambiguity (such as that between 'cistron' and 'recon') becomes troublesome, it is sometimes necessary to stipulate as clearly as possible which of the available options one is taking as a way of specifying the reference of one's terms. Even at the risk of total failure to refer—which might happen if one's analysis is mistaken—one fixes one's reference to all and only those things which fit a certain theoretical description. The result is clarity, and when clarity is the dominant intention, reference is fixed in much the way that Frege thought that it is always fixed. A sense is determined by a description, and reference depends on whether or not anything, in fact, fits that description. Finally, one may operate with a dominant intention which Kitcher calls *naturalism*, to wit, the intention to refer to the relevant effective natural kind occurring or operating in a certain situation or in a certain class of cases. Though the matter needs to be argued on another occasion, I suspect that one must have recourse to naturalism over and above conformity, accuracy, and clarity in order to put forth a successful account of the grounds on which Mendel, Bateson, Morgan, Benzer, and all the rest may be construed as employing concepts of the same thing—the gene.

Concluding Remarks

To set up the conclusion of this chapter, it will be useful to make some very small comments about the relationship between Mendelian and molecular genetics. As the reference of the term 'gene' became more tightly specified during the development of Mendelian (or, if you prefer, transmission) genetics, in a large range of central cases the concept of the gene became that of a minimal chromosomal segment performing a certain function or causing a certain effect. The relevant effect was known as the phenotype of the gene. Not surprisingly, a major part of the history of the gene, not addressed here, concerns the interplay between what one counts as genes and how one restricts or identifies the phenotypes which can be used to specify individual genes. But when all this is said and done, a great variety of phenotypes can legitimately be used to single out genes. As will soon be made clear, one's very concept of a gene depends on the range of phenotypes one considers.

Thanks to the advances made in molecular genetics, it is now possible to examine changes in the DNA (mutations!?!) fairly directly. In some cases, at least, it is also possible to track the effects of those changes rather exactly. It is now well known that some changes in the DNA are silent. That is, they have no effect on any other aspect of the structure, the development, or the composition of the organism. Effectively, such changes in the genetic material do not amount to changes in the function of any gene, though, when suitably located, they do constitute changes in the structure or composition of the relevant gene. Other changes in the DNA do, of course, change the organism, but some of them do so in ways which, arguably, are of no importance to its structure, development, or function. For example, some so-called point mutations result in the substitution of one amino acid for another in some particular protein manufactured in accordance with the information contained on the gene in question. Many such substitutions have very drastic effects. But some of them, so far as can be told, do not alter the way the protein folds and do not alter its biological activity or function in any significant way. In such cases, I suggest, there are strong reasons for tolerating in perpetuity important ambiguities in the concept of the gene.

The reason for this is that phenotypes at different levels are of concern for different purposes. Consider, for example, medical genetics. If one is concerned with PKU and allied metabolic disorders, the phenotypes one deals with will range from gross morphological and behavioral traits down to what turns out to be the heart of the matter—enzyme structure and function.[21] With respect to all of these phenotypes, both silent changes in the DNA and those changes which have no effect on enzyme structure or function will not (and should not) count as mutations, i.e., as changes in the relevant gene. It does not matter whether or not these changes occur within that segment of DNA which constitutes the gene of interest; because they have no relevant functional effects, the gene counts as unchanged. The reason for this is clear: the concept of the gene is coordinate with the concept of the phenotype. And the phenotype of concern is not defined biochemically at the level of DNA, but (if it is defined biochemically at all) at the level of protein or via some functional attributes consequent on the biochemistry of the relevant proteins.

It is important to recognize that there are legitimately different interests which lead us to deal with different sorts of phenotypes. Evolutionists, for example, may be interested in the rate of amino acid substitutions in proteins or of nucleotide substitutions in DNA. That is, the phenotypes they are concerned with might be defined by amino acid or even nucleotide sequence, not protein function. Accordingly, their definitions of the phenotype and of the gene may be discordant with those of the medical geneticist. And it is not a matter of right or wrong, but simply a matter of legitimately different interests. The point is a fairly deep one, allied to Putnam's point, discussed briefly above, about the linguistic division of labor. I have written, until now, sloppily, as if there were only one community of biologists, or, rather, geneticists. But this is simply false. There are large and important specialized subcommunities with legitimately different interests, interests which lead them to deal with legitimately different phenotypes. As the example introduced in this paragraph shows, there are serious cases in which there is no real question but that those differing phenotypes correspond with different concepts of the gene and different criteria for individuating genes.

Work in molecular genetics may well show that, like Bateson's, some contemporary attempts at establishing gene concepts are ill-founded.

Indeed, I believe that there are clear cases (for example in sociobiology, cf. Burian, 1981–82, but perhaps much more generally) in which certain gene concepts will simply have to be abandoned in light of some of the findings of molecular genetics. But molecular genetics cannot discriminate among well-founded gene concepts. There is a fact of the matter about the structure of DNA, but there is no single fact of the matter about what the gene is. Even though their concepts are discordant, the community of evolutionists concerned with the evolution of protein sequence and the community of medical geneticists working on metabolic disorders are both employing perfectly legitimate concepts of the gene. This provides strong, concrete support for the claim that the concept of the gene is open rather than closed with respect to both its reference potential and its reference.

A dangling thread provides a moral for biologists to consider. Recall Stadler's distinction between the operational concept of the gene and the various hypothetical concepts of the gene. Stadler is right that proper use of an operational concept can ensure conformity and protect against the pernicious effects of certain theoretical errors. But, as the example of white and eosin genes shows, operational criteria (here, specifically for the individuation of genes) are themselves theory-laden and quite often erroneous. Furthermore, there is no single operational concept (or set of operational criteria) for the gene. In the end, as the brief discussion of molecular genetics in the last few paragraphs suggests, the best arbiter we have of the legitimacy of both operational and hypothetical concepts of the gene comes from molecular analysis. The latter, in turn, cannot be extricated from what Stadler would have considered a hypothetical concept, namely that of the structure of the DNA molecule. It follows that genetic concepts (and theoretical concepts generally) are inescapably open in the ways I have been describing.

For philosophical readers, I add only that the results of this extremely sketchy treatment of the case of the gene, if they withstand scrutiny, will prove to be of immense consequence for our understanding of reduction and of scientific realism.

Notes

Previous versions of this chapter were read at The University of California, Davis; California State University, Fullerton; and Virginia Polytechnic Institute and State University. Comments on these occasions and by friends too numerous to mention have greatly improved the paper. I am grateful to all concerned.

1. The best source for this history is Carlson (1966), which in part inspired this essay. Cf. also Whitehouse (1965) for a complementary approach.

2. A useful elementary survey of the background is Brown (1977); a more advanced and detailed survey is Suppe (1977).

3. Cf. Feyerabend (1962, 1965, 1970, 1975).

4. Cf. Kuhn (1962, 1970, 1977).

5. Kuhn's original articulation of these ideas (1962, pp. 101–102) has been altered in various ways, but as his 1983a (pp. 566–567) shows, he still accepts the central claim that if the fundamental laws or postulates of a theory are not true, the leading theoretical terms of that theory do not refer. See the following note for an elaboration of this point.

6. Kuhn's latest attempt to escape the consequences of holism turns on restricting the interconnections among terms and concepts to a local context—hence his term *local holism*. Yet, as the following quotation shows, Kuhn has serious trouble in accounting for disagreements over the reference of theoretical terms: "Nevertheless, I take the Second Law to be necessary in the following language-relative sense: if the law fails, the Newtonian terms in its statement are shown not to refer." (Kuhn, 1983a, p. 567.)

   If this text is taken literally, Kuhn cannot properly parse a debate between an Einsteinian and a Newtonian physicist in which each individual argues that the theory he prefers gives a proper description of what, allegedly, they both refer to by the term 'mass'. This difficulty is a consequence of Kuhn's claim that if the Einsteinian term 'mass' is well-grounded, then the phonetically and lexicographically identical Newtonian term cannot refer at all. It follows that one cannot coherently argue that what is referred to by the Newtonian term is properly described within Einsteinian mechanics, for, while accepting Einstein's mechanics, one cannot coherently hold that the Newtonian term refers at all. Yet such disagreements are absolutely commonplace and give no sign of incoherence. What has gone wrong is the mistaken linkage of reference to meaning. Even if holism is correct about the meanings of theoretical terms (as it surely is), such holism is mistaken when applied to reference. This point undercuts all standard treatments of incommensurability, including the weakened versions which Kuhn currently advocates while seeking to escape the consequences of radical versions of incommensurability.

7. Cf., e.g., Nagel (1961). The extended debate among Hull (1974, 1976), Ruse (1976), Schaffner (1969, 1976), Wimsatt (1976, 1979), and others shows the sorts of difficulties which this approach encounters.

8. Useful secondary sources on Bateson include Cock (1983), Coleman (1970), and Darden (1977). For the invention of the term *genetics*, see Carlson (1966, pp. 15–16).

9. Cock (1983), who places considerably less emphasis than I do on Bateson's vibratory theory of heredity, has an extended discussion of Bateson's review of Morgan et al. at pp. 41ff.

10. My encounter with this notion in Kitcher (1978) was crucially formative; this chapter grew out of thinking through how to apply the apparatus developed there to the history of the concept of the gene. His development of this notion in his 1982, pp. 339–347, is particularly useful. I am also grateful to Kitcher for helpful discussions concerning our work in progress; the overlap of interest and approach is unusually strong.

11. A useful brief account of this theory may be found in Schwarz (1977).

12. The power of this position is nicely illustrated by Kuhn's tacit reliance on it in the argument criticized in n. 6. If if were not for the supposed unique connection between sense and reference, what ground would there be for holding that, given the way the world is, the reference of a theoretical term or concept is fully fixed by its place in the fundamental laws of that theory?

13. The point is entirely parallel with the objection to Kuhn put forward in n. 6.

14. Cf., e.g., Donnellan (1966, 1970), Kripke (1972), and Putnam, (1973, 1975).

15. Kitcher makes a similar point in different terminology in his 1982, p. 345.

16. Closure of reference with respect to theoretical terms is the fundamental move which undermines "local holism." So long as the fundamental laws of the relevant theory or the particular linguistic practices of a founding individual or community are thought to fix the reference of theoretical terms for once and for all, theory change which adjusts the reference of the resulting theoretical terms will be impossible, and referential change regarding theoretical entities will be an all or nothing phenomenon.

17. Among the modifications: Kitcher seems to think that reference potential is an extensional concept. On my reading, it is not. Thus, he claims to be an extensionalist. I am not.

18. Cf. Carlson (1966), chap. 8, for a discussion of the conceptual importance of Sturtevant's analysis which provided the key step in recognizing that mutation often involves alteration rather than loss of genes.

19. Cf. Benzer (1955, 1956, 1957).

20. Compare Kitcher's prior discussion in his 1982, pp. 342ff.

21. There is a helpful discussion of PKU in Burian (1981–82, pp. 55–59).

References

Bateson, W., 1916. [Review of Morgan, Sturtevant, Muller, and Bridges, *The Mechanism of Mendelian Heredity*]. *Science* 44:536–543.

Benzer, S., 1955. Fine structure of a genetic region in bacteriophage. *Proceedings of the National Academy of Science* 41:344–354.

Benzer, S., 1956. Genetic fine structure and its relation to the DNA molecule. *Brookhaven Symposia in Biology* 8:3–16.

Benzer, S., 1957. The Elementary Units of Heredity. In *The Chemical Basis of Heredity*, McElroy, W. D., and B. Glass, eds. Baltimore, Maryland: Johns Hopkins Press, pp. 70–93.

Brown, H. I., 1977. *Perception, Theory, and Commitment*. Chicago: Precedent Publishing.

Burian, R. M., 1981–1982. Human sociobiology and genetic determinism. *Philosophical Forum*: 43–66.

Carlson, E. A., 1966. *The Gene: A Critical History*. Philadelphia and London: W. B. Saunders Company.

Cock, A. G., 1983. William Bateson's rejection and eventual acceptance of chromosome theory. *Annals of Science* 40:19–59.

Coleman, W., 1970. Bateson and chromosomes: conservative thought in science. *Centaurus* 15:228–314.

Darden, L., 1977. William Bateson and the promise of mendelism. *Journal of the History of Biology* 10:87–106.

Donnellan, K., 1966. Reference and definite descriptions. *Philosophical Review* 75:281–304.

Donnellan, K., 1970. Proper names and identifying descriptions. *Synthese* 21:335–358.

Feyerabend, P., 1962. Explanation, reduction, and empiricism. In *Minnesota Studies in the Philosophy of Science*, vol. 3, Feigl, H., and G. Maxwell, eds. Minneapolis, Minnesota: University of Minnesota Press.

Feyerabend, P., 1965. Problems of Empiricism. In *Beyond the Edge of Certainty: University of Pittsburgh Series in the Philosophy of Science*, vol. 2. Colodny, R. G., ed. Englewood Cliffs, New Jersey: Prentice-Hall, pp. 145–260.

Feyerabend, P., 1970. Problems of Empiricism, Part II. In *The Nature and Function of Scientific Theories: Pittsburgh Series in Philosophy of Science*, vol. 4. Colodny, R. G., ed. Pittsburgh, Pennsylvania: University of Pittsburgh Press, pp. 275–353.

Feyerabend, P., 1975. *Against Method*. London: New Left Books.

Hull, D., 1974. *Philosophy of Biological Science*. Englewood Cliffs, New Jersey: Prentice-Hall.

Hull, D., 1976. Informal Aspects of Theory Reduction. In *PSA 1974*, R. S. Cohen, A. Hooker, A. C. Michalos, and J. W. Van Evra, eds. Dordrecht, Holland: Reidel, pp. 653–670.

Kitcher, P., 1978. Theories, theorists and theoretical change. *Philosophical Review* 87:519–547.

Kitcher P., 1982. Genes. *British Journal for the Philosophy of Science* 33:337–359.

Kripke, S., 1972. Naming and necessity. In *Semantics of Natural Language*, D. Dadson and G. Harman, eds. Dordrecht, Holland: Reidel, pp. 253–355.

Kuhn, T. S., 1962. *The Structure of Scientific Revolutions*. Chicago: University of Chicago Press.

Kuhn, T. S., 1970. Reflections on my critics. In *Criticism and the Growth of Knowledge*, I. Lakatos and A. Musgrave, eds. Cambridge: Cambridge University Press.

Kuhn, T. S., 1977. *The Essential Tension*. Chicago: University of Chicago Press.

Kuhn, T. S., 1983a. Commensurability, comparability, communicability. In *PSA 1982*, vol. 2, P. D. Asquith and T. Nickles, eds. East Lansing, Michigan: Philosophy of Science, pp. 669–688.

Kuhn, T. S., 1983b. Rationality and theory choice. *Journal of Philosophy* 80:563–570.

Morgan, T. H., and C. B. Bridges, 1913. Dilution effects and bicolorism in certain eye colors of *Drosophila*. *Journal of Experimental Zoology* 15:429–466.

Morgan, T. H., A. H. Sturtevant, H. J. Muller, and C. B. Bridges, 1915. *The Mechanism of Mendelian Heredity*. New York: Henry Holt and Co.

Nagel, E., 1961. *The Structure of Science*. New York: Harcourt, Brace and World.

Putnam, H., 1973. Explanation and reference. In *Conceptual change*, G. Pearce and P. Maynard, eds. Dordrecht, Holland: Reidel, pp. 199–221.

Putnam, H., 1975. The meaning of meaning. *Philosophical Papers*, vol. 2, Hilary, P., ed. Cambridge: Cambridge University Press.

Ruse, M., 1976. Reduction in genetics. In *PSA 1974*, R. S. Cohen, C. A. Hooker, A. C. Michalos, and J. W. Van Evra, eds. Dordrecht, Holland: Reidel, pp. 633–651.

Schaffner, K. F., 1969. The Watson-Crick model and Reductionism. *British Journal for the Philosophy of Science* 20:325–348.

Schaffner, K. F., 1976. Reductionism in biology. In *PSA 1974*, R. S. Cohen, C. A. Hooker, A. C. Michalos, and J. W. Van Evra, eds. Dordrecht, Holland: Reidel, pp. 613–632.

Schwarz, S. P., 1977. Introduction. In *Naming, Necessity, and Natural Kinds*, Schwarz, S. P., ed. Ithaca, New York: Cornell University press, pp. 13–41.

Stadler, L., 1954. The Gene. *Science* 120:811–819.

Sturtevant, A. H., 1913a. The Himalayan rabbit case, with some considerations on multiple allelomorphs. *American Naturalist* 47:234–238.

Sturtevant, A. H., 1913b. The linear arrangement of six sex-linked factors in *Drosophila*, as shown by their mode of association. *Journal of Experimental Zoology* 14:43–59.

Suppe, F., 1977. Introduction; Afterword—1977; and pp. 614–730. In *The Structure of Scientific Theories*, 2nd ed. Urbana, Illinois: University of Illinois Press, pp. 3–232, 617–730.

Watson, J. D., and F. H. C. Crick, 1953. Molecular structure of nucleic acids. *Nature* 171:737–738.

Whitehouse, H. L. K., 1965. *Towards an Understanding of the Mechanism of Heredity*. Stanton, England: Arnold.

Wimsatt, W., 1976. Reductive explanation: A functional account. In *PSA 1974*, R. S. Cohen, C. A. Hooker, A. C. Michalos, and J. W. Van Evra, eds. Dordrecht, Holland: Reidel, pp. 671–710.

Wimsatt, W. C., 1979. Reduction and reductionism. In *Current Problems in the Philosophy of Science*. H. Kyburg, Jr., and P. Asquith, eds. East Lansing, Michigan: Philosophy of Science Association, pp. 352–377.

How Biology Differs from the Physical Sciences

Ernst Mayr

For perhaps a hundred years there has been a strong tendency among philosophers to think that any science is like any other science and that physics could be accepted as the paradigm of science. This I consider to be a grievous error. I hope to be able to show that biology, even though as much a science as the physical sciences, is also in many ways very different. That there is a great difference has been almost completely ignored by all but the youngest generation of philosophers, and as a result when philosophers from Descartes to the logical positivists talked about science, they meant the physical sciences. I must have some six or seven volumes on my shelves claiming to be philosophies of science, but all of them deal exclusively with the physical sciences. Since many concepts that were developed in biology, particularly after 1859, are incompatible with the conceptual framework of the physical sciences, this inconvenient fact was eliminated by calling biology a "dirty science" or, as the physicist Ernest Rutherford said, as "postage stamp collecting." When the growing importance of biology was finally realized, a movement developed to restore the unity of science, but what was meant by this was to "reduce" biology to the physical sciences rather than to develop a broader philosophy of science that would do equal justice to the concepts of the physical and the life sciences.

This unfortunate situation is the product of history. When science reawakened after the Middle Ages and when the scientific revolution took place from Galileo to Newton and Lavoisier, it was the physical sciences that experienced this first flowering. Biology as a science was still dormant and did not really come to life until the 1830s or 1840s. For the philosophers, from Bacon, Descartes, and Locke to Kant, the physical sciences, and in particular mechanics, were the paradigm of science. This was so widely accepted that when biology finally developed in the nineteenth century, it tried to model itself as closely as possible on the example of physics. Kant, as has often been cited, claimed that any branch of knowledge contains only as much real science as it contains mathematics. And the kind of physics that appealed to the philosophers the most was classical, mechanical, deterministic physics in which everything obeyed universal laws, in which everything was due to movements and forces, and in which time did not exist. Essentialism ranked high as the philosophy of the physical sciences, chance was largely ignored, and experiment was considered the only acceptable scientific method. This caricature of physics was very convenient for exercises in logic and lent itself readily to the development of a philosophy of science dominated by essentialism, determinism, and reductionism.

I have consciously called my description of the thinking of the classical physical sciences a caricature, because it no longer fits the conceptual framework of modern physics. However, it was this thinking of classical mechanics that formed the original foundation of the philosophy of science, and it was this same extreme physicalism which led to the vitalistic backlash in biology. I am writing as a biologist and will not hold forth at length on the thinking of classical physics. However, I must point out a few more aspects of physicalism because this is needed for an understanding of the reaction by biologists.

The difference between the thinking of a physicist and a biologist can be conveniently illustrated by a few recent statements of the well-known physicist and Nobel laureate, Steven Weinberg. He said (1974, p. 56), "One of man's enduring hopes has been to find a few simple laws that would explain why nature with all of its seeming complexity and variety is the way it is." Surely no biologist would ever express such a hope. It would be difficult to expect that the incredible diversity of nature, the complexity of the process of ontogenetic differentiation and of the nervous system, or the qualitative uniqueness of each kind of macromolecule, could be expressed in the form of a "few simple general laws."

Another aspect of classical physicalism was its uncompromising reductionism. Here again we can quote Weinberg: "At the present moment the closest we can come to a unified view of nature is a description in terms of elementary particles and their mutual interactions." By contrast, every biologist would insist that to dissect complex biological systems into elementary particles would be by all odds the worst way to study nature. This would not further the understanding of nature in any way whatsoever except at the level of the atom and its components. At the risk of tossing out the baby with the bathwater, I am prepared to claim that, in view of the known frequency of repair mechanisms and feedback devices, disturbances at the level of elementary particles are ordinarily of no effect whatsoever at the higher levels of biological integration.

The classical philosophers of science tended to agree with the physicists that everything in the world of living organisms obeys the same laws as those that apply to inert matter and that there are no other laws. As a result there was an almost total neglect of specific biological phenomena and processes in the literature of the philosophy of science, a situation that lasted until about twenty-five years ago.

What was the reaction of biology to these claims? Broadly speaking two classes of responses can be distinguished. Perhaps the majority of the biologists, and particularly those working in physiology and other branches of functional biology, adopted the physicalist interpretation and attempted to explain all biological processes in terms of movements and forces. Everything was mechanistic, everything was deterministic, and there was no unexplained residue. As much as such authors as Helmholtz, Jacques Loeb, W. Roux, or Max Hartmann might otherwise differ from each other, their basic explanatory scheme was identical. Even naturalists like Weismann would glibly attribute differences among biological processes to differences in the "movement of their molecules." Jacques Loeb, Carl Ludwig, and Julius Sachs were perhaps the leaders of this physicalist biology. As productive as the physicalist approach was, particularly in physiology, it left vast areas of biology totally unexplained. The physicalists were like one of the blind men touching one part of an elephant.

But there were also some other biologists who were disturbed by unexplained aspects of biological processes. They felt, as had Aristotle more than two thousand years earlier, that a living organism had some sort of constituent that clearly distinguished it from inert matter. Separating this out less carefully than Aristotle had done, they called it by various names, such as *vital force, vis viva, Lebenskraft,* or *Entelechie.* This ingredient, they claimed, was outside the realm of the physical sciences and did not obey any of the laws of physics and chemistry. Those who had made these claims were called vitalists. For a vitalist, at least an extreme vitalist, there are two entirely separate worlds, that of the physical sciences and that of the world of life. Life and mind, the vitalists insisted, can only be explained by non-physical forces. In fact, interpretation in any branch of biology must take such vital forces into consideration, they said. The progressive trends in evolution, for instance, from the origin of life up to flowering plants, mammals, and man can only be accounted for if one postulates some orthogenetic force, an *élan vital* or *omega* principle, and such a force or principle was held to be outside the explanatory domain of chemistry and physics.

The leading vitalists in the eighteenth and nineteenth centuries were by no means ignorant metaphysicians but included some of the most intelligent and highly experienced biologists. Even in the first half of the twentieth century, those who adopted vitalism included some first-rate minds, such as Driesch and J. B. S. Haldane. Unfortunately, they were joined by a lot of lesser minds whose uninformed claims were nothing but damaging to any endeavor to establish an autonomous science of biology.

What is generally lumped under the term *vitalism* includes a heterogeneous mixture of theories of various degrees of validity. Delbrück (1971) demonstrated that for Aristotle vitalism was actually the postulate of a genetic program, as we would now call it, and when one carefully reads what Johannes Muller wrote about the *Lebenskraft* postulated by him, it is again a perfect description of the genetic program. There is little doubt that some of the much-maligned vitalists had a far more profound understanding of the living organism than their mechanistic opponents.

What is rather ironical is the fact that a number of physical scientists switched over to vitalism when they discovered that their naive Cartesian assumptions about the workings of organisms were wrong. It is well known that some of the leaders of quantum mechanics, such as Niels Bohr, Schroedinger, and Wolfgang Pauli postulated that some day one would discover unknown physical laws in organisms, laws not operating in inert matter. When Max Delbrück switched from physics to biology, it was in part in order to discover such laws if they existed.

Among biologists vitalism has been dead for some forty or fifty years. I was rather surprised when Francis Crick, as recently as 1966, found it necessary to write an entire book (*Molecules and Men*) that had the refutation of vitalism as its primary object. He justified this by saying that vitalism was not a matter of the past but that he had discovered three recent scientist-writers who had endorsed it. Who were these three recent alleged vitalists? To my amusement I discovered that all three of them were physical scientists.

The controversy between physicalists and vitalists continued for well over one hundred years. Both camps were convinced that they were right: the physicalists because no one was able to find even the slightest evidence for any force active in living organisms that did not obey the laws of physics and chemistry; and the vitalists because they could point to thousands of phenomena in the world of life which the crude laws of the physical sciences were unable to explain. How could this stalemate be broken?

By the 1930s or '40s it had become quite clear that neither of the two schools of biologists had found the right solution. Contrary to the claims of the physicalists, there were phenomena in living nature that were not explained by the reductionist physicalist approach. But at the same time it was equally clear that there were no processes, such as those postulated by the vitalists, that were not consistent with the laws of physics and chemistry. It became necessary to look for a third option, and the achievement of such an option was made possible by developments both in the physical sciences and in biology. Let me begin by describing some of the changes in the thinking of the physical scientists during the transformation from classical mechanics to modern physics.

The Erosion of Determinism

The thinking of physicists has changed rather drastically in the last one hundred years. As Prigogine has stated recently, "In the 19th century the physicists thought they had solved all the riddles of the universe. Newton's mechanics gave the answer to everything . . . now we know that we do not know the foundation of the world. Admitting this ignorance makes it much easier for physicists, biologists, and philosophers to come together" (1982, p. 124). I shall try to show in how many different ways the standards and criteria of the physical sciences have recently been relaxed and expanded in such a way as to facilitate the rapprochement of physics and biology.

What was perhaps the most characteristic feature of classical physics was its strict determinism. This is well illustrated by Laplace's famous (or infamous) statement. Biology, by contrast, is characterized by indeterminism and unpredictability. For instance, parents are never able to predict the sex of their next offspring, nor what particular mixture of the characteristics of the two parental families would determine its makeup. Some physiological processes, particularly molecular interactions, might have almost the predictability of a physical chemical process (not surprisingly, since this is what they essentially are), but there is always a strong indeterministic element either in the interaction of complex systems or in any more or less protracted sequence of events. At best the outcome can be predicted probabilistically. Traditionally one encountered statements, both in the literature of the physical sciences and in philosophy, that the physical sciences obey strictly deterministic laws, while biology, as J. Herschel said of evolutionary biology, obeyed the "law of the higgledy-piggledy." There seemed to be a total contrast between the two sciences.

All this has now changed and the absolute difference has been converted to one of degree. The physicists have learned this particularly when moving down to the level of elementary particles and to the development of quantum mechanics. This kind of uncertainty (Heisenberg's principle) is now so well known, as well as Bohr's complementarity principle, that I need say nothing further about it. What is often overlooked, however, is that one encounters even greater indeterminism when one moves from the study of unit processes, involving single entities, to that of larger systems. Whether one is dealing with ocean currents, weather systems, or in cosmology with nebulae and galaxies—all of them purely physical systems—everywhere one encounters strong turbulence, due to stochastic processes that preclude the making of strict deterministic predictions. Leading philosophers likewise have expressed vigorous opposition to determinism. I only have to mention Karl Popper's *The Open Universe: An Argument for Indeterminism*. In all, it is safe to state that it is no longer disreputable for a biologist to question the prevalence of determinism and of universal laws.

Laws and Other Notions

Laws

Laws are the cornerstone in the philosophy of the physical sciences. I shall not get myself entangled in the recent controversies concerning the relation between laws and causation. This much is certain: Until well into the twentieth century the physical sciences were dominated by a quest for laws. This greatly impressed biologists, who consequently, up to the end of the nineteenth century, also tried to explain all phenomena and processes as due to the operation of laws. In the *Origin of Species* Darwin refers to laws controlling certain biological processes no fewer than 106 times in 490 pages. One hundred years later, in my

Animal Species and Evolution (1963), which covers very much the same subject matter as Darwin's classic, I doubt that I invoked laws even a single time, except when quoting someone else. Indeed one will hardly ever encounter the word 'law' in any modern textbook of biology.

The working biologist sees two major problems with laws, which will become apparent when we scrutinize a customary definition of laws: "Laws are universal statements concerning classes of empirical entities and processes." The first problem concerns the qualification 'universal'. We find that with respect to more complex biological systems we can hardly ever make a generalization that does not have exceptions. To what extent is it permissible to recognize statistical or probabilistic laws? At what percentage of exceptions does a statistical law lose the right to be called a law? These are unanswered questions. Furthermore, what is the difference between a universal law and a simple fact? The statement All birds have feathers is a reference to a fact. What, if anything, is gained by calling it a law? One could ask the same question concerning many, and perhaps most, universal laws. They are simply facts.

Prediction

A belief in universal laws, of course, implies a belief in the possibility of absolute prediction. The goodness of prediction was, therefore, in classical physics considered the test for the goodness of an explanation. We biologists, I am afraid, have always been rather confused about the meaning of the word *prediction*. This is not surprising, considering the importance of time in most biological phenomena. In most cases, and particularly in evolutionary biology, a prediction meant to us a statement concerning an expected future event or condition. For instance: "No one at the beginning of the Cretaceous could have predicted that the dinosaurs would be extinct by the end of the geological period." Such a prediction obviously has nothing to do with the validity of a law. But even philosophers have sometimes been trapped by the ambiguity of the term prediction. At one time Karl Popper said something he would no longer say today: "Darwinism does not really predict the evolution of variety. It therefore cannot really explain it." As Scriven and other have shown, however, prediction is not a necessary part of causality. In complex systems, and in systems involving a great deal of stochastic perturbation, one can give a posteriori explanations of events which one could not have predicted with complete certainty.

Predictions in meteorology, cosmology, and other physical sciences that have to do with complex systems have exactly the same properties as those of the biological sciences. The possibility of making absolute predictions is no longer a distinguishing feature of the physical sciences as compared with biology.

Method

A similar relaxation of the formerly so rigid criteria of good science has occurred in many other respects. For instance, in the classical literature of the physical sciences it is stated again and again that experiment is the only method of science. Actually this has never been true. Copernicus and Kepler came to their conclusions on the basis of long-continued, painstaking observations of the movements of planets. And it was not only the observation of a single planet that led to the results, but also the comparison of the tracks of different planets. Observation and comparison have been among the most heuristic methods not only in astronomy but also in other physical sciences, such as geology, oceanography, and meteorology. Observation and comparison, of course, have always been of paramount importance in biology. This is as true for comparative anatomy or systematics as it is for evolutionary biology, ecology, and behavioral biology. Unfortunately observation and experiment are sometimes treated as opposites. Actually the experiment is very often the method by which conclusions based on observations are tested. On the other hand, there are distinct limitations to the domain of the experimental method, particularly when a historical dimension is involved. One cannot experiment with the history of stars nor test the various theories of the extinction of the dinosaurs by experiment. Likewise there are severe constraints on the utility and validity of computer simulations.

The most important conclusion one can draw from these considerations is that in principle there is no difference between the physical and the biological sciences with respect to experiment and observation. Both methods have their place in both sciences, and it depends on the particular field how important and useful either one or the other method is. In physiology and chemistry the experiment is predominant: in meteorology and evolutionary biology observation is the most important source of evidence that can be used for inferences.

Time

The original physical sciences were time-independent. History plays no role in the laws of physics. Hence when Darwin introduced time into the thinking of biology, this seemed to introduce a new contrast between the two fields. However, the more the physical sciences occupied themselves with systems, and particularly with complex systems as in cosmology and geology, the more it was realized that the past history of a system contributes to its characteristics. In astrophysics, in relativity theory, and even in small particle physics, thinking is now strongly influenced by considerations of time.

Systems

Theoretical physics tended to focus on the lowest components into which things could be reduced—molecules, atoms, and elementary particles. Molecules were considered to move independently and irregularly, as illustrated by Brownian movement. Now there is more and more emphasis on interactions, on evidence for self-organization in structures, patterns, and systems. All these have, of course, from the beginning, been concerns of biology.

Concepts

In modern biology there is a dominant concern with concepts—the development of new concepts and the modification of old ones. Concepts played a relatively minor role in classical physics. Karl Pearson, for instance, when describing the scope of science, indicates that he recognizes only facts and methods. "The unity of all science consists alone in its method, and not in its material" (p. 15). In addition to the method, he recognizes only facts and their classification. More recent authors increasingly emphasize the importance of themata (Holton) or of concepts.

I have presented this short discussion of recent changes in the physicists' thinking concerning determinism, laws, prediction, method, systems, and concepts in order to show that these profound changes require a rather drastic reorientation of the philosophy of physics. This will make it possible to develop a philosophy of science that is better qualified to do equal justice to both the physical and the biological sciences.

New Developments in Biology

The relation between physics and biology, however, has been affected not only by these changes in the thinking of physicists, but even more profoundly by an almost revolutionary change in the conceptual framework of biology. In the seventeenth and eighteenth centuries biology was largely physiology, while natural history was to such an extent allied to physico-theology that it was not seriously considered part of science. As far as the functional processes of physiology are concerned, it is to a large extent possible to reduce them to physico-chemical processes. An explanation of physiology in the standard terminology of the physical sciences—matter, movements, and forces—seemed indeed possible. Attempts to reduce all biological phenomena to the simplest laws and processes of physics continued through much of the nineteenth century. They are still promulgated by a few physicists, as in the recent statement of Francis Crick that "the ultimate aim of the modern movement in biology is in fact to explain *all* biology in terms of physics and chemistry" (Crick, 1966, p. 10).

In 1859 physiology lost in one stroke its position as the exclusive paradigm of biology, when Darwin established evolutionary biology. Even more importantly, biology thereby lost its unitary aspect. It took some time before this was fully realized, but eventually it became apparent that there are two biologies, the biology of proximate causations (functional biology) and that of ultimate causations (evolutionary biology) (Mayr, 1982, pp. 67–71). The biology of proximate causations deals with the functional processes of living organisms or, to put it in a different way, with the translation of genetic programs. Its major method is indeed the experiment, and its most important question is how. The biology of ultimate causations deals with evolutionary biology in the widest sense of the word. It occupies itself with the origin of new genetic programs, and its principal question is why. The two biologies require very different explanatory models, indeed very different philosophies. Functional biology, as we shall presently see, is not too far removed from the physical sciences, and is quite congenial with their methods and *Fragestellungen*; evolutionary biology, with its interest in historical processes, is in some respects as closely allied to the humanities as it is to the exact sciences.

As behavioral biology, ecology, biogeography, and other branches of modern biology developed, the image of science fashioned by physics, particularly by mechanics, became less and less suitable. The contrast between biology in this new broadened sense and the physical sciences became ever more apparent, and the estrangement between these sciences deepened. Yet, at the same time, further developments occurred in biology which prevented the conflict from becoming total. There were three developments in particular that laid the foundation for the ultimate reconciliation between the two sciences.

(1) The recognition by biologists that *all* processes in living organisms are consistent with the laws of physics and chemistry.

(2) A complete refutation of vitalism. As mentioned above, I know of no serious biologist who still maintains that there are forces or processes in organisms due to some specific life force or *Lebenskraft*. Some philosophers and physicists have attributed the recent autonomy movement of biology to a concealed return to vitalism. Nothing could be further from the truth. Every biologist is fully aware of the fact that molecular biology has demonstrated decisively that all processes in living organisms can be explained in terms of physics and chemistry. Vitalism is thus completely refuted.

(3) The realization that the differences that do exist between inanimate matter and living organisms are due to the organization of matter in living organisms.

The Emancipation of Biology

The question now before us is whether biology, after the refutation of vitalism, can be completely reduced to the laws and theories of the physical sciences. I will show that this is not the case. The physical sciences have to a considerable extent a logically unified structure, with a rather limited number of more or less universal laws. Most of these laws can be systematically related to each other. Nothing like this exists in biology. As I pointed out earlier, there are very few generalizations in biology that can be designated as laws. The question whether the absence, or at best the scarcity, of laws prevents biology from being considered a science, has recently been debated by philosophers. The claim was made by Smart (1963, pp. 52, 57) that biology is nothing but a sophisticated kind of natural history. Smart's analysis of biology has been severely criticized, because a branch of science does not need to have a structure of universal laws in order to qualify as science. Michael Ruse (1970; 1973, pp. 24–31), Michael Simon (1971, pp. 9–21), and Ronald Munson (1975, pp. 444–445) have shown that any biological generalization that has no reference to any particular species or particular gene, etc. is spatio-temporally unrestricted, and hence is as general as the laws of the physical sciences. However, most of these generalizations can be formulated only in probabilistic terms. This is true even for such a comprehensive and all-pervasive principle as natural selection. Perhaps one can call the so-called Central Dogma of molecular biology a law. This states that proteins cannot be translated back into nucleic acids, and that an inheritance of acquired characters is therefore impossible. Most biologists, however, consider this finding simply a fact. The same is true for the theory of common descent, which can be stated as a law in the form All organisms have descended from common ancestors. Again: is this a law or simply a fact?

A number of authors have attempted to construct a theoretical biology based on an underlying mathematical structure. None of these attempts has been successful. Most of what is most characteristic of living organisms cannot be expressed in mathematical terms or in terms of the simplistic laws of physics. All this has led to the inevitable conclusion that physicalism is in every respect as unsuitable as a basis for a philosophy of biology as vitalism.

As is always the case in intractable scientific controversies, it has become clear that a third option must be found which, in its explanatory model, avoids the weaknesses of both physicalism and vitalism. Such a new approach would have to be based on an understanding of the special characteristics of organisms, and their contrast to the structure of inanimate matter. This includes such insights as that systems, and particularly ordered systems, are not the same as singular entities; that there is a hierarchy of systems, with the properties of the higher systems not necessarily reducible to those of the lower ones; and that biological systems store historically acquired information. These are only a few of the characteristics we shall presently discuss in detail. They all document the fact that one cannot reduce biological phenomena and processes to purely physical ones. It is this conclusion that gives validity to the call for an autonomous biology.

A plea for the autonomy of biology does not mean that the unity of science is thereby undermined. Rather, it signifies a demand that the concept of science be placed on a broader foundation by abandoning the narrow constraints of the physical sciences. No one has stated this more forcefully than George Gaylord Simpson (1964, pp. 106–107):

Insistence that the study of organisms requires principles additional to those of the physical sciences does not imply a dualistic or vitalistic view of nature. Life . . . is not thereby necessarily considered as non-physical or nonmaterial. It is just that living things have been affected for . . . billions of years by historical processes. . . . The results of those processes are systems different in kind from any nonliving systems and almost incomparably more complicated. They are not for that reason necessarily less material or less physical in nature. The point is that *all* known material processes and explanatory principles apply to organisms, while only a limited number of them apply to nonliving systems. Biology, then, is the science that stands at the center of all science . . . and it is here, in the field where all the principles of all the sciences are embodied that science can truly become unified.

Special Aspects of the World of Life

These are rather sweeping claims. To what extent are they justified? This can be determined only by enumerating the principles and concepts that pertain exclusively to the world of life but not to that of inanimate objects. Perhaps most conspicuous among these, but perhaps philosophically least important, is the difference in the substance of living organisms. Their chemical uniqueness consists in the fact that they are built of high-molecular-weight macromolecules not found in inanimate nature. To be sure, some of the lower molecular weight components of these macromolecules, for instance amino acids or purines and pyrimidines, are found in the cosmic dust, but the higher molecular compounds are inseparately connected with life. And, of course, how they were once formed is the great problem of the origin of life.

One can make long lists of the characterisitics of life and of living processes, but in order to avoid technicalities and to facilitate discussion I shall assemble them under three major headings.

Uniqueness and Variability

Nothing is perhaps more characteristic of the world of living organisms than the universality of uniqueness. No two individuals in a sexually reproducing species are identical. Among the millions of cells of an organisms no two are probably exactly identical, owing to the diverse activity (suppression and activation) of regulatory genes. Even greater, of course, is uniqueness among species, higher taxa, and ecosystems. Uniqueness, to be sure, is not altogether absent in the inanimate world either, but it is restricted to highly complex systems such as stars, weather systems, ocean currents, and mountains. The basic entities in the physical sciences, elementary particles, atoms, and molecules, not only form large classes of identical objects, but are also highly constant in contrast to living systems like cells or individuals, which change continuously. Uniqueness results in variability, and variability is characteristic of living systems from the cells of the body through individuals to species and to still higher aggregations. It is this variability that necessitated a shift in the philosophical foundations, a shift from *typological essentialism* to *population thinking*. What do I mean by these terms?

Western thinking for the more than 2,000 years after Plato was dominated by essentialism. This is the term given by Karl Popper to that part of Plato's philosophy that deals with the *eidos*. Essentialism is vividly characterized by Plato's allegory of the shadows on the cave wall. The vast world of variable phenomena in the universe consisted for Plato of imperfect reflections of a fixed number of constant, discontinuous *eidē*. It was not until the nineteenth century that a new type of thinking, population thinking, emerged (Mayr, 1982, pp. 45–47). For the population thinker the variation of the population, determined by the uniqueness of the constituent individuals, is the reality of nature, and mean values are only statistical abstractions.

Philosophy is to a large extent still dominated by typological essentialism. Replacing it by population thinking changes many of our conclusions. The whole theory of paradoxes, as accepted by logicians, is based on essentialism. The theory of natural selection is incomprehensible to an essentialist, as was very evident in the opposition to natural selection in the post-Darwinian decades. As recently as 1967 it was apparent that a group of physicists and mathematicians, thinking in terms of typological essentialism, was unable to comprehend the possibility of a simultaneous variation of thousands of gene loci (Moorhead and Kaplan, 1967) and were thus unable to make a correct determination of the rate of evolution.

The consequences arising from uniqueness are many. It explains the almost incomprehensible diversity of the living world. It explains why in the course of evolution so often different organisms adopt different pathways to achieve the same adaptation (the principle of multiple pathways). It explains why the response to a selection pressure is only probabilistic. Indeed, it is one of the reasons why predictions in biology are so often impossible. We will not have an acceptable philosophy of science until all the consequences of the uniqueness principle in the living world are properly accommodated by such a philosophy.

Systems and Their Hierarchical Organization

The second great peculiarity of living organisms is that they are highly complex, ordered systems. Complexity per se is not a fundamental difference between organic and inorganic systems. The world weather system or any galaxy are also complex systems. But on average, systems in the world of organisms are by several orders of magnitude more complex than those of inanimate objects. I have recently described this as follows:

Complexity in living systems exists at every level from the nucleus (with its DNA program), to the cell, to any organ system (like kidney, liver, or brain), to the individual, the ecosystem, or the society. Living systems are invariably characterized by elaborate feedback mechanisms, unknown in their precision and complexity in any inanimate system. They have the capacity to respond to external stimuli, the capacity for metabolism (binding or release of energy), and the capacity to grow and to differentiate (Mayr, 1982, p. 53).

Furthermore, living systems do not have a random complexity but are highly organized. All these characteristics are used to define life.

Complex systems usually have a hierarchical structure, the entities of one level being compounded into new entities at the next higher level, as cells into tissues, tissues into organs, and organs into functional systems. To be sure, hierarchical organization is also found in the inanimate world, such as between elementary particles, atoms, molecules, crystals, and so on; but it is in living systems that hierarchical structure is of special significance (Mayr, 1982, pp. 64–66).

The predominant role played by organized, highly integrated systems in biological research is a further difference between the physical and the biological sciences. When one asks where the greatest gaps in our understanding of organisms are, three sets of problems are usually mentioned: (1) the control of differentiation during ontogeny; (2) the workings of the central nervous system; and (3) the interaction of controlling factors in ecosystems.

In each case one is dealing with enormously complex systems, with a very high number of interacting components, regulatory mechanisms, and feedbacks. Whenever it is possible to isolate single components and unitary processes from such systems, it is found that they are completely explicable by known chemico-physical laws. What is unknown and unpredictable, however, in most of these cases is the regulation of the interactions of the vast number of components. Nowhere in the inanimate world can one find a system, even a complex system, that has the ordered internal cohesion and coadaptation of even the simplest of biological systems. And this requires an entirely different approach from that of the classical philosophy of science.

Systems at each hierarchical level have two characteristics. They act as wholes (as if they were a homogenous entity), and their characteristics cannot (not even in theory) be deduced from the most complete knowledge of the components, taken separately or in other partial combinations. In other words, when such systems are assembled from their components, new characteristics of the new whole emerge that could not have been predicted from a knowledge of the components. Such emergence is quite universal, occurring also, of course, in inanimate systems, but it nowhere else plays the important role that it does in living organisms. Perhaps the two most interesting characteristics of new wholes are that they can in turn become parts of still higher-level systems, and that they can affect properties of components at lower levels (downward causation) (Campbell, 1974, p. 182).

The belief that wholes may be more than the sum of their parts has also been designated *holism*. When first proposed, this had a vitalistic flavor, but the modern emergentist accepts constitutive reduction without reservation, and this fully excludes vitalism. Nor does belief in emergence mean that the organism can be studied only as a whole. On the contrary, every effort must be made to carry the analysis of its components, and of their components, as far as this is possible, always realizing, however, that this will not necessarily explain the emergent qualities of such systems at higher hierarchical levels.

Recognition of the importance of emergence demonstrates, of course, the invalidity of extreme reductionism. By the time we have dissected an organism down to atoms and elementary particles we have lost everything that is characteristic of a living system.

For a long time philosophers were skeptical of the validity of the principle of emergence. Even though acceptance of this principle was argued ably by Lewes (1874–75) and Lloyd Morgan (1894), it is only in recent decades that its acceptance has become widespread among philosophers. As Karl Popper said, "We live in a universe of emergent novelty" (1974, p. 281). It is now perfectly obvious that biological phenomena and processes cannot be completely reduced to processes found in inanimate matter. This again justifies the validity of a call for an autonomous biology.

Possession of a Genetic Program

Now let me take up the third of the great peculiarities of living organisms. It is the fact that there is one difference between them and inanimate matter that is absolute. All organisms possess a historically evolved genetic program, coded in the DNA of the nucleus of the zygote (or in RNA in some viruses). Nothing comparable exists in the inanimate world, except for manmade computers. The presence of this program gives organisms a peculiar duality, consisting of a phenotype and a genotype. The genotype (unchanged except for occasional mutations) is handed on from generation to generation. In interaction with the environment it controls the production of the phenotype, that is, of the visible organism that we encounter and study.

A number of the unique characteristics of the genetic program must be emphasized. One of its properties is that it can supervise its own precise replication, as well as that of other living systems such as organelles, cells, and whole organisms. There is nothing exactly equivalent in all of inorganic nature. Another important aspect of the genotype is that it is the result of a history that goes back to the origin of life, and thus incorporates the "experiences" of all ancestors (Delbrück, 1949). Finally, it endows organisms with the capacity for teleonomic (goal-directed) processes and activities, a capacity totally absent in the inanimate world. Except for the twilight zone of the origin of life, the possession of a genetic program provides for an absolute difference between organisms and inanimate matter.

Since each genotype, that is, each genetic program, is a unique combination of thousands of different genes, the difference cannot be expressed in quantitative, but only in qualitative terms. Thus quality becomes one of the dominant aspects of living organisms and their characteristics. This is particularly obvious in a comparison of courtship displays, of pheromones, of niche occupations, and almost any characteristic of a particular species of organism.

Many activities of organisms, particularly of animals, are designated *behavior*. Behavior involves dynamic processes of a nature unknown in the inanimate world. Perhaps the two most important characteristics, from the point of view of philosophy, are, first, that behavior is adaptive—in other words that it facilitates survival and reproductive success and permits a fine-grained adaptation to the environment—and, second, that changes of behavior very often function as pacemakers in evolution, by leading organisms into new niches or environments, which exert a new set of selection pressures and thus may lead to major evolutionary changes.

A Philosophy of Science That Includes Biology

What consequences arise for philosophy by recognizing the autonomy of biology? It seems to me that the answer to this question might be the following. By expanding the concept of science so as to include biology in all of its aspects, it is possible to construct a philosophy that is far richer and far more suited for man than a philosophy largely based on the physical sciences. There is no pathway from the laws of physics to man. The more physicists could explain by Newtonian laws and more powerful theories of relativity and quantum mechanics, the more they isolated man in the universe. As a result, as Heisenberg has stated, it has happened "that for the first time in the course of history, man on earth faces only himself, that he finds no longer any other partner or foe" (Heisenberg, 1958, pp. 103–105). No naturalist/biologist has ever had that feeling, however. Ever since 1859 biologists have felt themselves to be a part of nature, inanimate as well as living nature. And if one is a truly thinking biologist, one has a feeling of responsibility for nature, as reflected by much of the conservation movement.

It is high time that the consequences of the Darwinian revolution be fully incorporated into philosophy. It is time that it be realized how much more important the Darwinian revolution was than any recent revolutions in the physical sciences. The changes in the theories of physics brought about by Maxwell, Planck, Einstein, Heisenberg, or Schrödinger have no effect on the personal philosophy of the man in the street. It is quite different with the Darwinian revolution, which profoundly altered everybody's view of nature and of himself. This is the reason a philosophy of science based exclusively on the principles of the physical sciences is insufficient and incomplete. What we need is the development of a broadly based philosophy of science that incorporates not only the concepts of the physical but also those of the biological sciences. Such a broadened philosophy must not only acknowledge the important differences between the physical and biological sciences but also, and more importantly, it must acknowledge that the principles and concepts of the biological sciences are not opposed to those of the physical sciences, but are supplementary. This is the reason it is so important that the findings of biology, in all their consequences, should be built into philosophy. Karl Popper has clearly realized this and has expressed it in a series of his most recent writings (Popper, 1974; 1981).

It is sometimes asked whether biology is creating a new, post-mechanistic world view (Lewontin, 1983). If I had been asked this question, I would have said that I could not answer it, unless I was told what is meant by the term *mechanistic world view*. If mechanistic is defined as the opposite of vitalistic then I would have answered no: modern biology recognizes the occurrence of no processes in living organisms that are in conflict with or outside the laws of physics and chemistry.

However, if mechanistic were defined in terms of simpleminded mechanics, and more specifically in a reductionist-deterministic manner, then I would say that biology indeed is creating a world view that is in conflict with any physicalist world view that ignores all that is characteristic of the world of life and that ignores everything not encountered in the world of inanimate objects.

On what insights should a new philosophy of science be based if it is to fully reflect the new understanding of living nature? Let me list, by way of summation, some of the more important concepts that must be duly considered:

• that the historical nature of organisms is acknowledged, in particular their possession of a historically acquired genetic program;

• that individuals at most hierarchical levels, from the cell up, are unique and form populations whose variance is one of their major characteristics;

• that there are two biologies, functional biology, which asks proximate questions, and evolutionary biology, which asks ultimate questions;

• that the patterned complexity of living systems is hierarchically organized and that the higher levels in this hierarchy are characterized by the emergence of novelties;

• that observation and comparison are methods in biological research that are fully as scientific and heuristic as the experiment;

• that an insistence on the autonomy of biology does not mean an endorsement of vitalism, orthogenesis, or any other theory that is in conflict with the laws of chemistry or physics; and

• that a philosophy of science must include a consideration of the major biological concepts not only of molecular biology and physiology but also of evolutionary biology, systematics, behavioral biology, and ecology.

Biology belongs to science just as much as do physics and chemistry. There is no hope for a truly comprehensive philosophy of science until the autonomous features of living organisms are truly recognized. A unification of the sciences, likewise, cannot be achieved by simply ignoring the special attributes of the world of life, but rather by recognizing them clearly and by including them appropriately in the design of the conceptual structure of science as a whole. This is the reason I emphasize so strongly the features by which living organisms differ from inanimate matter.

References

Campbell, D., 1974. 'Downward causation' in hierarchically organized biological systems. In *Studies in the Philosophy of Biology*, F. J. Ayala and Th. Dobzhansky, eds. Berkeley and Los Angeles: University of California Press.

Crick, F., 1966. *Of Molecules and Men*. Seattle: University of Washington Press.

Delbrück, M., 1949. A physicist looks at biology. *Trans. Conn. Acad. Arts. Sci.* 38:173–190.

Delbrück, M., 1971. Aristotle-totle-totle. In *Of Microbes and Life*, J. Monod and E. Borek, eds. New York: Columbia University Press, pp. 50–55.

Heisenberg, W., 1958. *Philosophy: The Revolution in Modern Science*. New York: Harper and Row.

Lewontin, R. C., The corpse in the elevator, *New York Review of Books*, Jan. 20, 1983, pp. 34–37.

Mayr, E., 1982. *The Growth of Biological Thought*. Cambridge: Belknap Press of Harvard University Press.

Moorhead, P. S., and M. M. Kaplan, eds., 1967. *Mathematical Challenges to the Neo-Darwinian Interpretation of Evolution*. Wistar Institute Symposium Monograph, No. 5. Philadelphia: Wistar Institute Press.

Munson, R., 1975. Is biology a provincial science? *Philos. Sci.* 42:428–447.

Pearson, K., 1982. *The Grammar of Science*. London: Walter Scott.

Popper, K. R., 1974. Scientific reduction and the essential incompleteness of all science. In *Studies in the Philosophy of Biology*, F. J. Ayala and Th. Dobzhansky, eds. Berkeley and Los Angeles: University of California Press, pp. 259–284.

Popper, K. R., 1981. *The Open Universe. An Argument for Indeterminism*. London: Hutchinson.

Prigogine, I., 1982. Es gibt keine wirkliche Evolution, wenn alles gegeben ist. *Dialektik im Gesprach* 5:121–133.

Ruse, M. E., 1970. Are there laws in biology? *Australasian Journ. Phil.* 48:234–246.

Simon, M., 1971. *The Matter of Life*. New Haven: Yale University Press.

Simpson, G. G., 1964. *This View of Life*. New York: Harcourt, Brace and World.

Smart, J. J. C., 1963. *Philosophy and Scientific Realism*. London: Routledge and Kegan Paul.

Weinberg, S., 1974. Unified theories of elementary-particle interaction. *Scientific American* 231:56.

Reduction in Biology:
A Recent Challenge

Francisco J. Ayala

Scientific theories have sometimes been shown to be interconnected by establishing that the principles of a theory or branch of science can be explained by the principles of another theory or branch of science of greater generality. The less general theory, called the secondary theory, is then said to have been reduced to the more general or primary theory. The integration of diverse scientific theories into more comprehensive ones simplifies science and expands the explanatory power of scientific principles. Thus it conforms to the goals of science.

The reduction of a theory or even a whole branch of science to another has occurred repeatedly in the history of science (Nagel, 1961; Popper, 1974). One impressive example is the reduction of thermodynamics to statistical mechanics, made possible by the discovery that the temperature of a gas reflects the mean kinetic energy of its molecules. Several branches of physics and astronomy have been to a large extent unified by their reduction to a few theories of great generality, such as quantum mechanics and relativity. A large sector of chemistry became reduced to physics after it was shown that the valence of an element bears a simple relation to the number of electrons in the outer orbit of the atom.

In biology, parts of genetics have been to some extent reduced to chemistry after discovery of the structure and mode of replication and action of the hereditary material, DNA. More generally, the impressive success of molecular biology has prompted some authors to claim that the goal of biology is to account for biological phenomena in terms of the underlying physico-chemical process; i.e., that mature biology should be but a specialized branch of the physical sciences. Whether biology can be reduced to physical science is a question that has received much attention in recent decades (see Koestler and Smythies, 1969; Ayala and Dobzhansky, 1974). I have explored the matter elsewhere

(Ayala, 1968, 1977) and need not repeat here my analysis. Recently, however, the issue of reductionism in biology has been posed in another context, namely, the relationship between the study of micro- and macroevolution. It is this that I propose here to examine.

Some paleontologists (Gould, 1980, 1982a,b; Stanley, 1979, 1982, Vrba, 1980) have recently argued that macroevolution—the evolution of species, genera, and higher taxa—is an autonomous field of study, irreducible to microevolutionary theory. This claim to autonomy has been expressed as a "decoupling" of macroevolution from microevolution (Stanley, 1979, pp. x, 187, 193), as a rejection of the notion that microevolutionary mechanisms can be extrapolated to explain macroevolutionary processes (Gould, 1980, p. 383), or as an explicit argument against the reductionism of macroevolution to microevolution (Gould, 1982a, p. 94).

The paleontologists who argue for the autonomy of macroevolution base their claim on the notion that large-scale evolution is "punctuated," rather than "gradual." The model of punctuated equilibrium proposes that morphological evolution happens in bursts, with most phenotypic change occurring during speciation events, so that new species are morphologically quite distinct from their ancestors, but do not thereafter "change substantially in phenotype over a lifetime that may encompass many million years." The punctuational model is contrasted with the gradualistic model, for which morphological change is a more or less gradual process. Such change largely occurs during the lifetime of a species, and is not strongly associated with speciation events.

Two quotations should suffice to establish that punctuationalists predicate the autonomy of macroevolution on the alleged punctuational nature of large-scale evolution. "If rapidly divergent speciation interposes discontinuities between rather stable entities (lineages), and if there is a strong random element in the origin of these discontinuities (in speciation), then phyletic trends are essentially *decoupled* from phyletic trends within lineages. Macroevolution is decoupled from microevolution" (Stanley, 1979, p. 187, italics in the original). "Punctuated equilibrium is crucial to the independence of macroevolution—for it embodies the claim that species are legitimate individuals, and therefore capable of displaying irreducible properties" (Gould, 1982, p. 94).

Whether phenotypic change in macroevolution occurs in bursts or is more or less gradual, is a question to be decided empirically. Examples of rapid phenotypic evolution followed by long periods of morphological stasis are known in the fossil record. But there are instances as well in which phenotypic evolution appears to occur gradually within a lineage. The question is the relative frequency of one or the other mode; and paleontologists disagree in their interpretation of the fossil record. (Eldredge, 1971; Eldredge and Gould, 1972; Hallam, 1978; Raup, 1978; Stanley, 1979; Gould, 1980; and Vrba, 1980, are among those who favor punctualism; whereas Kellog, 1975; Gingerich, 1976; Levinton and Simon, 1980; Schopf, 1979 and 1981; Cronin, Boaz, Stringer, and Rak, 1981; and Douglas and Avise, 1982, favor phyletic gradualism.)

Whatever the paleontological record may show about the frequency of smooth, relative to jerky, evolutionary patterns, there is, however, one fundamental reservation that must be raised against the theory of punctuated equilibrium. This evolutionary model argues not only that most morphological change occurs in rapid bursts followed by long periods of phenotypic stability, but also that the bursts of change occur during the origin of new species. Stanley (1979, 1982), Gould (1982a,b), and other punctuationalists have made it clear that what is distinctive in the theory of punctuated equilibrium is this association between phenotypic change and speciation. For example: "Punctuated equilibrium is a specific claim about speciation and its deployment in geological time; *it should not be used as a synonym for any theory of rapid evolutionary change at any scale.* . . . Punctuated equilibrium holds that accumulated speciation is the root of most major evolutionary change, and that what we have called anagenesis is usually no more than repeated cladogenesis (branching) filtered through the net of differential success at the species level" (Gould, 1982a, pp. 84–85; italics added).

Species (in sexually reproducing organisms) are groups of interbreeding natural populations that are reproductively isolated from any other such groups (Mayr, 1963; Dobzhansky et al., 1977). Speciation involves, by definition, the development of reproductive isolation between populations previously sharing in a common gene pool. But it is in no way apparent how the fossil record could provide evidence of the development of reproductive isolation. Paleontologists recognize species by their different morphologies as preserved in the fossil record. New species that are morphologically indistinguishable from their ancestors (or from contemporary species) go totally unrecognized. Sibling species are common in many groups of insects, in rodents, and in other well-studied organisms (Mayr, 1963; Dobzhansky, 1970; Nevo and Shaw, 1972; Dobzhansky et al., 1977; White, 1978; Benado, Aguilera, Reig, and Ayala, 1979). Moreover, morphological discontinuities in a time series of fossils are usually interpreted by paleontologists as speciation events, even though they may represent phyletic evolution within an established lineage, without splitting of lineages.

Thus, when paleontologists use evidence of rapid phenotypic change in favor of the punctuational model, they are guilty of definitional circularity. Speciation as seen by the paleontologist always involves substantial morphological change—and morphological change always happens in association with speciation—because paleontologists identify new species by the eventuation of substantial morphological change.

Three Distinct Issues

According to the proponents of punctuated equilibrium, phyletic evolution proceeds at two levels. First, there is change within a population that is continuous through time. This consists largely in allelic substitutions prompted by natural selection, mutation, genetic drift, and the other processes familiar to the population geneticist, operating at the level of the individual organism. The proponents of punctuated equilibrium argue that neo-Darwinism or the Synthetic Theory of Evolution sees evolution mostly as a process of allelic substitutions through time. Yet they claim that most evolution within established lineages rarely, if ever, yields any substantial morphological change. Second, there is the process of origination and extinction of species. Most morphological change is associated with the origin of new species. Evolutionary trends result from the patterns of origination and extinction of species, rather than from evolution within established lineages. Hence, the relevant unit of macroevolutionary study is the species rather than the individual organism. It follows from this argument that the study of microevolutionary processes provides little, if any, information about macroevolutionary patterns, the tempo and mode of large-scale evolution. Thus macroevolution is autonomous relative to microevolution, in much the same way that biology is autonomous relative to physics. Gould (1982b, p. 384) has summarized the argument: "Individuation of higher-level units is enough to invalidate the reductionism of traditional Darwinism—for pattern and style of evolution depend critically on the disposition of higher-level individuals [i.e., species]."

As so often happens with questions of reductionism, the issue of whether microevolutionary mechanisms can account for macroevolutionary processes is muddled by confusion of separate issues. Identification of the issues involved is necessary in order to resolve them and to avoid misunderstanding, exaggerated claims, or unwarranted fears.

The issue whether the mechanisms underlying microevolution can be extrapolated to macroevolution involves, at least, three separate questions: (1) whether microevolutionary processes operate (and have operated in the past) throughout the organisms which make up the taxa in which macroevolutionary phenomena are observed; (2) whether the microevolutionary processes identified by population geneticists (mutation, random drift, natural selection) are sufficient to account for the morphological changes and other macroevolutionary phenomena observed in higher taxa, or whether additional microevolutionary processes need to be postulated; (3) whether theories concerning evolutionary trends and other macroevolutionary patterns can be derived from knowledge of microevolutionary processes.

The distinctions that I have made may perhaps become clearer if I state them as they might be formulated by a biologist concerned with the question whether the laws of physics and chemistry can be extrapolated to biology. The first question would be whether the laws of physics and chemistry apply to the atoms and molecules present in living organisms. The second question would be whether interactions between atoms and molecules according to the laws known to physics and chemistry are sufficient to account for biological phenomena, or whether the workings of organisms require additional kinds of interactions between atoms and molecules. The third question would be whether biological theories can be derived from the laws and theories of physics and chemistry.

Do Microevolutionary Processes Underlie Macroevolutionary Phenomena? The first issue can easily be resolved. It is unlikely that any biologist would seriously argue that the laws of physics and chemistry do not apply to the atoms and molecules that make up living things. Similarly, it seems unlikely that any paleontologist or macroevolutionist would claim that mutation, drift, natural selection, and other microevolutionary processes do not apply to the organisms and populations which make up the higher taxa studied in macroevolution. There is, of course, an added snarl—macroevolution is largely concerned with phenomena of the past. Direct observation of microevolutionary processes in populations of long-extinct organisms is not possible. But there is no reason to doubt that the genetic structures of populations living in the past were in any fundamental way different from the genetic structures of living populations. Nor is there any reason to believe that the processes of mutation, random drift and natural selection, or the nature of the interactions between organisms and the environment would have been different in nature for, say, Paleozoic trilobites or Mesozoic ammonites than for modern molluscs or fishes. Extinct and living populations—like different living populations—may have experienced quantitative differences in the relative importance of one or another process, but the processes could have hardly been different in kind. Not only are there reasons to the contrary lacking, but the study of biochemical evolution also reveals a remarkable continuity and gradual change of informational macromolecules (DNA and proteins) over the most diverse organisms, which advocates that the current processes of population change have persisted over evolutionary history (Dobzhansky et al., 1977).

Can Microevolutionary Processes Account for Macroevolution? The second question is considerably more interesting than the first: Can the microevolutionary processes studied by population geneticists account for macroevolutionary phenomena or do we need to postulate new kinds of genetic processes? The large morphological (phenotypic) changes observed in evolutionary history, and the rapidity with which they appear in the geological record, is one major matter of concern. Another issue is "stasis," the apparent persistence of species, with little or no morphological change, for hundreds of thousands or millions of years. The dilemma is that microevolutionary processes apparently yield small but continuous changes, while macroevolution as seen by punctuationalists occurs by large and rapid bursts of change followed by long periods without change.

Goldschmidt (1940, p. 183) argued long ago that the incompatibility is real: "The decisive step in evolution, the first step towards macroevolution, the step from one species to another, requires another evolutionary method than that of sheer accumulation of micromutations." Goldschmidt's solution was to postulate "systemic mutations," yielding "hopeful monsters" that, on occasion, would find a new niche or way of life for which they would be eminently preadapted. The progressive understanding of the nature and organization of the genetic material acquired during the last forty years excludes the "systemic mutations" postulated by Goldschmidt, which would involve transformations of the genome as a whole.

Some single-gene or chromosome mutations are known to have large effects on the phenotype because they act early in the embryo and their effects become magnified through development. Examples of such macromutations carefully analyzed in *Drosophila* are bithorax and the homeotic mutants that transform one body structure (antennae) into another (legs). Whether the kinds of morphological differences that characterize different taxa are due to such macromutations or to the accumulation of several mutations with small effect, has been examined particularly in plants where fertile interspecific, and even intergeneric, hybrids can be obtained. The results do not support the hypothesis that the establishment of macromutations is necessary for divergence at the macroevolutionary level (Stebbins, 1950; Clausen, 1951; Grant, 1971; see Stebbins and Ayala, 1981). The same conclusion has been reached by Lande (1981 and references therein; see also Charlesworth et al., 1982), who has convincingly shown that major morphological changes, such as in the number of digits or limbs, can occur in a geologically rapid fashion through the accumulation of mutations each with a small effect. The analysis of progenies from crosses between races or species that differ greatly (by as much as 30 phenotypic standard deviations) in a quantitative trait indicates that these extreme differences can be caused by the cumulative effects of no more than 5 to 10 independently segregating genes.

The punctuationalist claim that mutations with large phenotypic effects must have been largely responsible for macroevolutionary change is based on the rapidity with which morphological discontinuities appear in the fossil record (Stanley, 1979; Gould, 1980). But the alleged evidence does not necessarily support the claim. Microevolutionists and macroevolutionists use different time scales. Events that appear instantaneous in the geological time scale may involve thousands, even millions of generations. Gould (1982a, p. 84), for example, has made operational the fuzzy expression "geologically instantaneous" by suggesting that "it be defined as 1 percent or less of later existence in stasis. This permits up to 100,000 years for the origin of a species with a subsequent life span of 10 million years." But 100,000 years encompasses one million generations of an insect such as *Drosophila*, and tens or hundreds of thousands of generations of fish, birds, or mammals. There is little doubt that the gradual accumulation of small mutations may yield sizeable morphological changes during periods of that length.

Anderson's (1973) study of body size in *Drosophila pseudoobscura* provides an estimate of the rates of gradual morphological change produced by natural selection. Large populations, derived from a single set of parents, were set up at different temperatures and allowed to evolve on their own. A gradual, genetically determined change in body size ensued, with flies kept at lower temperatures becoming, as expected, larger than those kept at higher temperatures. After 12 years, the mean size of the flies from the population kept at 16°C had become, when tested under standard conditions, approximately 10 percent greater than the size of the flies kept at 27°C. The change of mean value was greater than the standard deviation in size at the time when the tests were made. Assuming 10 generations per year, the populations diverged at a rate of 8×10^{-4} of the mean value per generation.

Paleontologists have emphasized the "extraordinary high *net* rate of evolution that is the hallmark of human phylogeny" (Stanley, 1979). Interpreted in terms of the punctuationalist hypothesis, human phylogeny would have occurred as a succession of jumps, or geologically instantaneous saltations, interspersed with long periods without morphological change. Could these bursts of phenotypic evolution be due to the gradual accumulation of small changes? Consider cranial capacity, the character undergoing the greatest relative amount of change. The fastest rate of net change occurred between 500,000 years ago, when our ancestors were represented by *Homo erectus* and 75,000 years ago, when Neanderthal man had acquired a cranial capacity similar to that of modern humans. In the intervening 425,000 years, cranial capacity evolved from about 900 cc in Peking man to about 1400 cc in Neanderthal people. Let us assume that the increase in brain size occurred in a single burst at the rate observed in *D. pseudoobscura* of 8×10^{-4} of the mean value per generation. The change from 900 cc to 1400 cc could have taken place in 540 generations or, assuming a generous 25 years per generation, in 13,500 years. Thirteen thousand years are, of course, a geological instant. Yet this evolutionary "burst" could have taken place by gradual accumulation of mutations with small effects at rates compatible with those observed in microevolutionary studies.

The known processes of microevolution can, then, account for macroevolutionary change, even when this occurs according to the punctuationalist model—i.e., at fast rates concentrated on geologically brief time intervals. But what about the problem of stasis? The theory of punctuated equilibrium argues that after the initial burst of morphological change associated with their origin, "species generally do not change substantially in phenotype over a lifetime that may encompass many million years" (Gould, 1982b, p. 383). Is it necessary to postulate new processes, as yet unknown to population genetics, in order to account for the long persistence of lineages without apparent phenotypic change? The answer is no.

The geological persistence of lineages without morphological change was already known to Darwin, who wrote in the last edition of *The Origin of Species* (1872, p. 375). "Many species once formed never undergo any further change. . .; and the periods, during which species have undergone modification, though long as measured by years, have probably been short in comparison with the periods during which they retain the same form." A successful morphology may remain unchanged for extremely long periods of time, even through successive speciation events—as manifested, for example, by the existence of sibling species that in many known instances have persisted for millions of years (Stebbins and Ayala, 1981).

Evolutionists have long been aware of paleontological stasis and have explored a number of alternative hypotheses consistent with microevolutionary principles and sufficient to account for the phenomenon. Although the issue is far from definitely settled, the weight of the evidence favors stabilizing selection as the primary process responsible for morphological stasis of lineages through geological time (Stebbins and Ayala, 1981; Charlesworth et al., 1982).

The Epistemological Autonomy of Macroevolution. Macroevolution and microevolution are not decoupled in the two senses so far expounded: identity at the level of events and compatibility of theories. First, the populations in which macroevolutionary patterns are observed are the same populations that evolve at the microevolutionary level. Second, macroevolutionary phenomena can be accounted for as the result of known microevolutionary processes. That is, the theory of punctuated equilibrium (as well as the theory of phyletic gradualism) is consistent with the theory of population genetics. Indeed, any theory of macroevolution that is correct must be compatible with the theory of population genetics, to the extent that this is a well-established theory.

Now, I pose the third question raised earlier: can macroevolutionary theory be derived from microevolutionary knowledge? The answer can only be no. If macroevolutionary theory were deducible from microevolutionary principles, it would be possible to decide between competing macroevolutionary models simply by examining the logical implications of microevolutionary theory. But the theory of population genetics is compatible with both punctuationalism and gradualism; hence logically it entails neither. Whether the tempo and mode of evolution occur predominantly according to the model of punctuated equilibria or according to the model of phyletic gradualism is an issue to be decided by studying macroevolutionary patterns, not by inference from microevolutionary processes. In other words, macroevolutionary theories are not reducible (at least at the present state of knowledge) to microevolution. Hence macroevolution and microevolution are decoupled in the sense (which is epistemologically most important) that macroevolution is an autonomous field of study that must develop and test its own theories.

Hierarchy and Emergence: Invalid Grounds for Epistemological Autonomy

Punctuationalists have claimed autonomy for macroevolution because species—the units studied in macroevolution—are higher in the hierarchy of organization of the living world than individual organisms. Species, they argue, have therefore "emergent" properties, not exhibited by, nor predictable from, lower-level entities. In Gould's (1980, p. 121) words, the study of evolution embodies "a concept of hierarchy—a world constructed not as a smooth and seamless continuum, permitting simple extrapolation from the lowest level to the highest, but as a series of ascending levels, each bound to the one below it in some ways and independent in others . . . 'emergent' features not implicit in the operation of processes at lower levels, may control events at higher levels." Although I agree with the thesis that macroevolutionary theories are not reducible to microevolutionary principles, I shall argue that it is a mistake to ground this autonomy on the hierarchical organization of life, or on purported emergent properties exhibited by higher-level units.

The world of life is hierarchically structured. There is a hierarchy of levels that runs from atoms, through molecules, organelles, cells, tissues, organs, multicellular individuals and populations, to communities. Time adds another dimension to the evolutionary hierarchy, with the interesting consequence that transitions from one level to another occur: as time proceeds the descendants of a single species may include separate species, genera, families, etc. But hierarchical differentiation of subject matter is neither necessary nor sufficient for the autonomy of scientific disciplines. It is not necessary, because entities of a given hierarchical level can be the subject of diversified disciplines: cells are appropriate subjects of study for cytology, genetics, immunology, and so on. Even a single event can be the subject matter of several disciplines. My writing of this paragraph can be studied by a physiologist interested in the workings of muscles and nerves, by a psychologist concerned

with thought processes, by a philosopher interested in the epistemological question at issue, and so on. Nor is the hierarchical differentiation
of subject matter a sufficient condition for the autonomy of scientific
disciplines: relativity theory obtains all the way from subatomic particles
to planetary motions and genetic laws apply to multicellular organisms
as well as to cellular and even subcellular entities.

One alleged reason for the theoretical independence of levels within
a hierarchy is the appearance of "emergent" properties, which are "not
implicit in the operation of lesser levels," and "may control events at
higher levels." The question of emergence is an old one, particularly
in discussions of the reducibility of biology to the physical sciences.
The issue is, for example, whether the functional properties of the
kidney are simply the properties of the chemical constituents of that
organ. In the context of macroevolution, the question is: do species
exhibit properties different from those of the individual organisms of
which they consist? I have argued elsewhere (Dobzhansky et al., 1977,
ch. 16) that questions about the emergence of properties are ill-formed,
or at least unproductive, because they can only be solved by definition.
The proper way of formulating questions about the relationship between
complex systems and their component parts is by asking whether the
properties of complex systems can be inferred from knowledge of the
properties that their components have in isolation. The issue of emergence cannot be settled by discussions about the nature of things or
their properties, but it is resolvable by reference to our knowledge of
those objects.

Consider the following question: Are the properties of common salt,
sodium chloride, simply the properties of sodium and chlorine when
they are associated according to the formula NaCl? If among the properties of sodium and chlorine I include their association in table salt
and the properties of the latter, the answer is yes; otherwise, the answer
is no. But the solution, then, is simply a matter of definition; and
resolving the issue by a definitional maneuver contributes little to
understanding the relationships between complex systems and their
parts.

Is there a rule by which one could decide whether the properties of
complex systems should be listed among the properties of their component parts? Assume that by studying the components in isolation
we can infer the properties they will have when combined with other
component parts in certain ways. In such a case, it would seem reasonable to include the "emergent" properties of the whole among the

properties of the component parts. (Notice that this solution to the problem implies that a feature that may seem emergent at a certain time, might not appear as emergent any longer at a more advanced state of knowledge.) Often, no matter how exhaustively an object is studied in isolation, there is no way to ascertain the properties it will have in association with other objects. We cannot infer the properties of ethyl alcohol, proteins, or human beings from the study of hydrogen, and thus it makes no good sense to list their properties among those of hydrogen. The important point, however, is that the issue of emergent properties is spurious and that it needs to be reformulated in terms of propositions expressing our knowledge. It is a legitimate question to ask whether the statements concerning the properties of organisms (but not the properties themselves) can be logically deduced from statements concerning the properties of their physical components.

Conditions for Epistemological Reduction

The question of the autonomy of macroevolution, like other questions of reduction, can only be settled by empirical investigation of the logical consequences of propositions, and not by discussions about the nature of things or their properties. What is at issue is not whether the living world is hierarchically organized—it is; or whether higher level entities have emergent properties—which is a spurious question. The issue is whether in a particular case, a set of propositions formulated in a defined field of knowledge (macroevolution) can be derived from another set of propositions (microevolutionary theory). Scientific theories consist, indeed, of propositions about the natural world. Only the investigation of the logical relations between propositions can establish whether one theory or branch of science is reducible to some other theory or branch of science. This implies that a discipline that is autonomous at a given stage of knowledge may become reducible to another discipline at a later time. The reduction of thermodynamics to statistical mechanics became possible only after it was discovered that the temperature of a gas bears a simple relationship to the mean kinetic energy of its molecules. The reduction of genetics to chemistry could not take place before the discovery of the chemical nature of the hereditary material. (I am not, of course, intimating that genetics can now be fully reduced to chemistry, but only that a partial reduction may be possible now, whereas it was not before the discovery of the structure and mode of replication of DNA.)

Nagel (1961; see also Ayala, 1968) has formulated the two conditions that are necessary and, jointly, sufficient to effect the reduction of one theory or branch of science to another. These are the condition of derivability and the condition of connectability.

The condition of derivability requires that the laws and theories of the branch of science to be reduced be derived as logical consequences from the laws and theories of some other branch of science. The condition of connectability requires that the distinctive terms of the branch of science to be reduced be redefined in the language of the branch of science to which it is reduced—this redefinition of terms is, of course, necessary in order to analyze the logical connections between the theories of the two branches of science.

Microevolutionary processes, as currently known, are compatible with the two models of macroevolution—punctuationalism and gradualism. From microevolutionary knowledge, we cannot infer which one of those two macroevolutionary patterns prevails, nor can we deduce answers for many other distinctive macroevolutionary issues, such as rates of morphological evolution, patterns of species extinctions, and historical factors regulating taxonomic diversity. The condition of derivability is not satisfied: the theories, models, and laws of macroevolution cannot be logically derived, at least at the present state of knowledge, from the theories and laws of population biology.

In conclusion, then, macroevolutionary processes are underlain by microevolutionary phenomena and are compatible with microevolutionary theories, but macroevolutionary studies require the formulation of autonomous hypotheses and models (which must be tested using macroevolutionary evidence). In this (epistemologically) very important sense, macroevolution is decoupled from microevolution: macroevolution is an autonomous field of evolutionary study.

References

Anderson, W. W., 1973. Genetic divergence in body size among experimental populations of *Drosophila pseudoobscura* kept at different temperatures. *Evolution* 27:278–284.

Ayala, F. J., 1968. Biology as an autonomous science. *American Scientist* 56:207–221.

Ayala, F. J., 1977. Philosophical Issues. In *Evolution*, Th. Dobzhansky, F. J. Ayala, G. L. Stebbins, and J. W. Valentine. San Francisco: W. H. Freeman, pp. 474–516.

Ayala, F. J., and Th. Dobzhansky, eds., 1974. *Studies in the Philosophy of Biology*. London: Macmillan; and Berkeley: University of California Press.

Benado, M., M. Aguilera, D. A. Reig, and F. J. Ayala, 1979. Biochemical genetics of Venezuelan spiny rats of the Proechimys guaine and Prochimys trinitatis superspecies. *Genetica*59:89–97.

Charlesworth, B., R. Lande, and M. Slatkin, 1982. A neo-Darwinian commentary on macroevolution. *Evolution* 36:474–498.

Clausen, J. 1951. *Stages in the Evolution of Plant Species*. Ithaca: Cornell University Press.

Cronin, J. E., N. T. Boaz, C. B. Stringer, and Y. Rak, 1981. Tempo and mode in hominid evolution. *Nature* 292:113–122.

Dobzhansky, Th., 1970. *Genetics of the Evolutionary Process*. New York: Columbia University Press.

Dobzhansky, Th., F. J. Ayala, G. L. Stebbins, and J. W. Valentine, 1977. *Evolution* 36:224–232.

Douglas, M. E., and J. C. Avise, 1982. Speciation rates and morphological divergence in fishes: Tests of gradual versus rectangular modes of evolutionary change. *Evolution* 36:224–232.

Eldredge, N., 1971. The allopatric model and phylogeny in paleozoic invertebrates. *Evolution* 25:156–167.

Eldredge, N., and S. J. Gould, 1972. Punctuated equilibria: an alternative to phyletic gradualism. In *Models in Paleobiology*, T. J. M. Schopf, ed. San Francisco: W. H. Freeman, pp. 82–115.

Gingerich, P. D., 1976. Paleontology and phylogeny: Patterns of evolution at the species level in early tertiary mammals. *American J. Science* 276:1–28.

Goldschmidt, R. B., 1940. *The Material Basis of Evolution*. New Haven: Yale University Press.

Gould, S. J., 1980. Is a new and general theory of evolution emerging? *Paleobiology* 6:119–130.

Gould, S. J., 1982a. The meaning of punctuated equilibrium and its role in validating a hierarchical approach to macroevolution. In *Perspectives in Evolution*, R. Milkman, ed. Sunderland, Massachusetts: Sinauer Assoc.

Gould, S. J., 1982b. Darwinism and the expansion of evolutionary theory. *Science* 216:380–387.

Grant, V., 1971. *Plant Speciation*. New York: Columbia University Press.

Hallam, A., 1978. How rare is phyletic gradualism and what is its evolutionary significance? Evidence from Jurassic bivalves. *Paleobiology* 4:16–25.

Kellogg, D. E., 1975. The role of phyletic change in the evolution of *Pseudocubus vema* (Radiolaria). *Paleobiology* 1:359–370.

Koestler, A., and J. R. Smythies, 1969. *Beyond Reductionism*. London: Hutchinson.

Lande, R., 1981. The minimum number of genes contributing to quantitative variation between and within populations. *Genetics* 99:541–553.

Levinton, J. S., and C. M. Simon, 1980. A critique of the punctuated equilibria model and implications for the detection of speciation in the fossil record. *Systematic Zoology* 29:130–142.

Mayr, E., 1963. *Animal Species and Evolution*. Cambridge, Massachusetts: Harvard University Press.

Nagel, E., 1961. *The Structure of Science*. New York: Harcourt, Brace and World, Inc.

Nevo, E., and C. R. Shaw, 1972. Genetic variation in a subterranean mammal, *Spalax ehrenbergi*. *Biochemical Genetics* 7:235–241.

Popper, K. R., 1974. Scientific reduction and the essential incompleteness of all science. In *Studies in the Philosophy of Biology*, F. J. Ayala and Th. Dobzhansky, eds. London: Macmillan.

Raup, D. M., 1978. Cohort analysis of generic survivorship. *Paleobiology* 4:1–15.

Schopf, T. J. M., 1979. Evolving paleontological views on deterministic and stochastic approaches. *Paleobiology* 5:337–352.

Schopf, T. J. M., 1981. Punctuated equilibrium and evolutionary stasis. *Paleobiology* 7(2):156–166.

Stanley, S. M., 1979. *Macroevolution: Pattern and Process*. San Francisco: W. H. Freeman.

Stanley, S. M., 1982. Macroevolution and the fossil record. *Evolution* 36:460–473.

Stebbins, G. L., 1950. *Variation and Evolution in Plants*. New York: Columbia University Press.

Stebbins, G. L., and F. J. Ayala, 1981. Is a new evolutionary synthesis necessary? *Science* 213:967–971.

Vrba, E. S., 1980. Evolution, species, and fossils: How does life evolve? *South African Journal Science* 76:61–84.

White, M. J. D., 1978. *Modes of Speciation*. San Francisco: W. H. Freeman.

Adaptation Explanations:
Are Adaptations for the Good of Replicators or Interactors?

Robert N. Brandon

In a recent paper Richard Dawkins asks, "When we say that an adaptation is 'for the good of' something, what is that something?" (1982a, p. 47). Dawkins's answer is, I think, surprising. If right, it should occasion a radical shift in our conception of nature. But I will argue that his answer is incorrect. In so doing, I hope to shed new light on the nature and value of adaptation-explanations in biology.

Evolutionary explanations of adaptations have received considerable attention from both biologists and philosophers of science. Many authors, including the present one, have described such explanations as teleological.[1] Others, primarily biologists, have argued that the term *teleology* should be dropped from the biologist's vocabulary; in its stead the term *teleonomy* has been introduced.[2] Notwithstanding this semantic controversy, there is wide agreement among neo-Darwinists that the study and explanation of biological adaptations is one of the central problems of evolutionary biology (the other traditionally being the explanation of the origin and diversity of species).

In an earlier paper (Brandon, 1981) I tried to explicate precisely the sense in which evolutionary explanations of adaptations are teleological. My views on that have not changed. I still think that the term *teleological* is perfectly appropriate in describing such explanations. Indeed it is more than appropriate, for it properly emphasizes the important similarities to certain pre-Darwinian (most notably, Aristotelian) explanations of adaptations. But, cognizant as I am of the allergic reaction the term produces in most biologists, I will refrain from using it in this chapter.

What is the connection between adaptation explanations and Dawkins's question? To claim that something is an adaptation is, I will argue, to make a claim about the causal history of that thing and, so to say, how its existence is to be causally explained. In a sense, adaptations are explained in terms of the good they confer on some entities. (This will be made more precise later.) Dawkins's question then concerns the nature of the entities benefited.

Replicators and Interactors

An important distinction in neo-Darwinism is that between genotype and phenotype. According to the received view, natural selection acts on phenotypes, not on genes or genotypes (Mayr, 1963). For instance, if there is selection for increased height in a population, then the taller organisms will tend to leave more offspring than their shorter con- specifics. This could be due to their living longer, having a higher reproductive rate, or whatever. In any case a necessary condition for such selection to occur is that there be phenotypic variation; if all the organisms within a population are of the same height, there can be no selection for increased height.

Natural selection alone is not sufficient to produce evolutionary change—the trait under selection must also be heritable. In our example taller-than-average parents must tend to produce taller-than-average offspring, and shorter-than-average parents shorter-than-average off- spring. But, of course, height is not directly transmitted from parent to offspring; rather, genes are. If height is genetically heritable, then off- spring of taller-than-average parents will tend to have genotypes dif- ferent from those of offspring of shorter-than-average parents. In the process of ontogeny these genotypic differences will manifest themselves as phenotypic differences. Thus, to quote Mayr, evolution by natural selection is "a two-step process" (1978). Natural selection acts on phe- notypic variation (produced by genotypic variation) which results in the differential replication of genotypes.

Although this traditional description seems wholly adequate for cases of natural selection occurring at the level of organismic phenotypes, it lacks generality. It cannot be applied easily to selection occurring at other levels, for instance gametic, group, or species levels. David Hull (1980 and 1981) and Richard Dawkins (1982a and 1982b) have therefore introduced a distinction between replicators and interactors which is best seen as a generalization of the traditional genotype-phenotype distinction.[3]

Dawkins defines a *replicator* as "anything in the universe of which copies are made" (1982a, p. 83). Genes are paradigm examples of replicators, but this definition does not preclude other things from being replicators. For instance, in asexual organisms the entire genome would be a replicator, and in cultural evolution ideas—or what Dawkins calls *memes*—may be replicators (Dawkins, 1976).

In any case, the qualities making for good replicators are longevity, fecundity, and fidelity (Dawkins, 1978). Here longevity means longevity in the form of copies. It is highly unlikely that any particular DNA molecule will live longer than the organism in which it is housed. What is of evolutionary importance is that it produce copies of itself so that it is potentially immortal in the form of copies. Of course, everything else being equal, the more copies a replicator produces (fecundity) and the more accurately it produces them (fidelity), the greater its longevity and evolutionary success.

In explicating Dawkins's notion of replicators, Hull stresses the importance of directness of replication. Although according to Dawkins organisms are not replicators, they may be said to produce copies of themselves. This replication process may not be as accurate as that of DNA replication, but nonetheless there is a commonality of structure produced through descent from parent to offspring. However, there is an important difference in the directness of replication between these two processes. The height of a parent is not directly transmitted to its offspring. As discussed earlier, that transmission proceeds indirectly through genic transmission and ontogeny. In contrast, genes replicate themselves less circuitously. Both germ-line replication (meiosis) and soma-line replication (mitosis) are physically quite direct. The importance of this is made explicit in Hull's definition. He defines replicators as "entities which pass on their structure directly in replication" (1981, p. 33).

Dawkins offers a fourfold classification of replicators: they may be active or passive, and (cutting across that classification) they may be germ-line or dead-end (1982, p. 83). An active replicator is one that influences its probability of being replicated. For example, any DNA molecule which, either through protein synthesis or the regulation of protein synthesis, has some phenotypic effect is an active replicator. A passive replicator is one which has no influence on its probability of being copied. A section of DNA that is never transcribed (so-called selfish DNA; Doolittle and Sapienza, 1980; Orgel and Crick, 1980; or Dawkins, 1982b, chapter 9) may be a passive replicator.[4]

Replicators, whether active or passive, are classified as germ-line or dead-end depending on whether they are potential ancestors of an indefinitely long line of descendant replicators. A gene in a gamete or, as the name implies, in a germ-line cell in a body is a germ-line replicator. Most of the genes in our bodies are not germ-line and can replicate only a finite number of times (through mitosis). These are dead-end replicators. One should note that it is the potential, not the fact, of being in an indefinitely long ancestry that matters for this classificatory distinction. So a gene in a spermatozoon that fails to fertilize an egg is still a germ-line replicator.

As discussed above, evolution by natural selection is a two-step process. One step involves the direct replication of structure. The other involves some interaction with the environment so that replication is differential. The entities functioning in the latter step have traditionally been called phenotypes. Although this notion seems adequate for discussions of organismic selection, entities other than organisms (e.g., chromosomes, groups of organisms or, perhaps, entire species) can interact with their environment in a way that makes replication differential. Thus, again in the interest of generality, Hull (1980, 1981) suggests the term *interactor* and defines it as "an entity that directly interacts as a cohesive whole with its environment in such a way that replication is differential" (1980, p. 318).

Although Hull and Dawkins are largely in agreement concerning the replicator-interactor distinction, there are two differences worth noting. The first is purely terminological. Dawkins has not adopted Hull's term interactor; instead he uses the term *vehicle*. According to Dawkins a vehicle is "any relatively discrete entity, such as an individual organism, which houses replicators . . . and which can be regarded as a machine programmed to preserve and propagate the replicators that ride inside it" (1982b, p. 295). On the whole I prefer Hull's term and his definition of it, and so in this paper I will use the term interactor. The term vehicle, by conjuring up images of little genes riding around in splendid isolation, seems to prejudice the case as to what sorts of entities are of primary importance in accounting for evolution by natural selection.

The second difference is more substantive. Dawkins holds that any change in replicator structure is passed on in the process of replication (1982b, p. 85 or 1982a, p. 51). Thus, given the truth of Weismannism (the doctrine that there is a one-way causal influence from germ line to body), replicators are importantly different from most interactors (e.g., organisms). But it is easy to imagine self-repairing replicators, where changes in structure would not necessarily be passed on in replication. What seems to be important is that replication be direct and accurate. But both *directness* and *accuracy* are terms of degree, and if we allow some play in both, then under certain circumstances an organism could be considered a replicator (which would not preclude its being an interactor as well). For example, Hull argues that:

> . . . one paramecium can divide into two, each new individual possessing the same structure as the original. In such cases, Dawkins would say that the entire genome is functioning as a replicator because its structure remains intact in the process. The justification which I am suggesting concerns the structure of the organism itself. But a quite similar situation can occur in sexual reproduction. When two organisms mate which are genetically homozygous and quite similar to each other, the resulting offspring are likely to be equally similar, regardless of how much recombination occurs. Once again, Dawkins would say that the genomes are functioning as replicators, while I would argue that the organisms themselves might be viewed as replicators, the only difference being in how direct the mechanism of replication happens to be (1981, p. 34).

I will not try to resolve this dispute here, but it will be relevant in the next section.

Having reviewed the replicator-interactor distinction we can now return to the question with which this paper began. When we speak of an adaptation as being "for the good of something," what is that something? That something is, according to Dawkins, the unit of natural selection or the *optimon* (1982b, p. 81). Using Dawkins's term, our question then becomes: What sort of thing is an optimon?

Dawkins argues that optimons are active germ-line replicators. The following, I think, fairly summarizes his argument:

> The reason active germ-line replicators are important units is that, wherever in the universe they may be found, they are likely to become the basis for natural selection and hence evolution. If replicators exist that are active, variants of them with certain phenotypic effects tend to out-replicate those with other phenotypic effects. If they are also germ-line replicators, these changes in relative frequency can have long-term, evolutionary impact. The world tends automatically to become populated by germ-line replicators whose active phenotypic effects are such as to ensure their successful replication. It is these phenotypic effects that we see as adaptations to survival. When we ask *whose* survival they are adapted to ensure, the fundamental answer has to be not the group, nor the individual organism, but the relevant replicators themselves (1982, p. 84).

Adaptation Explanations and the Units of Selection

Dawkins argues that the ultimate units of selection, that is, the units ultimately benefited by adaptations, are active germ-line replicators.

In this section I want to argue against that position and argue that interactors are the ultimate units of selection (in the above sense).[5] Before doing that, we will have to discuss briefly (1) the nature of adaptation-explanations; (2) the sense (if there is one) in which they legitimize talk of adaptations as "for the good of something"; and (3) the question of units and levels of selection.

In an earlier paper (Brandon, 1981a) I have given an account of the structure of adaptation-explanations. Here I will only briefly summarize that account. But first I should say something about my basic approach. Explanations are best understood as answers to certain sorts of questions. Some scientific questions call for explanations as answers; others don't. For instance, What is the mass of the sun? and At present in the United States, to what extent is height genetically heritable in humans? do not. But How does a warbler navigate when migrating? and Why is the sun spheroidal? do. (In English how and why are fairly reliable indicators of questions calling for explanations, and what is a decent indicator of questions not calling for explanations as answers. But the distinction is not syntactic.) If explanations are answers to certain sorts of questions, then important differences between types of explanations may well be discerned by looking at differences in the types of questions to which they serve as answers. This approach to the study of explanations is termed *erotetic*, that is, one based on the logic of the questions to be answered.

I am not the first person to suggest an erotetic approach to explanations. The approach goes back at least to Aristotle.[6] More recently, Mayr distinguishes functional biology (and the explanations offered by that branch of biology) from evolutionary biology on the basis of differences in the questions asked. He writes:

The functional biologist is vitally concerned with the operation and interaction of structural elements, from molecules up to organs and whole individuals. His ever-repeated question is "How?" How does something operate, how does it function? . . . The evolutionary biologist differs in his method and in the problems in which he is interested. His basic question is "Why?" (1961, p. 1502).

Finally, the erotetic approach has its followers among contemporary philosophers.[7] But one should note how radically it differs from the positivist approach that has dominated this century. That approach, usually associated with Hempel, ignores and obscures the relation between questions and explanations (Hempel, 1965).

In the earlier paper already mentioned (Brandon, 1981a) I argued that adaptation-explanations in biology should be distinguished from other evolutionary explanations (both in and out of biology) on the basis of the former but not the latter being answers to what-for questions. Questions concerning putative adaptations, an anteater's tongue, the structure of the human eye, or the waggle-dance of honeybees—are naturally formulated using what-for. (One might also ask the same questions using why or how-come. The distinction is not a simple syntactic one.) In contrast, we balk at using what-for in formulating other evolutionary questions, such as Why is hydrogen more abundant in the universe than uranium? I argued that our differentiation of these two sorts of evolutionary questions is quite proper and is founded on important differences between physical evolution and evolution by natural selection (Brandon, 1981a, pp. 95–97).

For present purposes we need only concern ourselves with adaptation-explanations. Having characterized them as answers to what-for questions, we need to delimit the proper objects of such questions. Not surprisingly, they are adaptations. But what is an adaptation?

The question is a complicated one, and fully justifying my answer to it would require more space than I have; so again I will try briefly to summarize what I have discussed more fully elsewhere (Brandon, 1981a, pp. 97–102).

Working backward, let me first state my answer. An adaptation is a phenotypic trait, or to use our present terminology, a feature of an interactor that has evolved as a direct product of natural selection. What is natural selection? It is differential reproduction of interactors that is due to differential adaptedness to a common environment. Finally, adaptedness is the complex dispositional property of interactors, the differences in which explain nonrandom differential reproduction (Brandon, 1978 and 1981b).

According to the above, to say that a trait is an adaptation is to make a claim about its causal history. It is also to claim that the trait evolved as a direct product of natural selection. It is not, I hasten to add, a theorem of current evolutionary theory that all organismic traits (or traits of any other sort of interactor) have the requisite causal history to be adaptations. Traits may appear and may even evolve in a population without being adaptations. Without trying exhaustively to cat-alogue the types of traits that are not adaptations, let me mention three important ones. First, there are traits due to chance. For example, a trait may appear based on a new single mutation. It would be placed in our first category regardless of its effect on the adaptedness of its possessor. (If it increased the adaptedness of its possessors, and because of that increased in relative frequency over generational time, then it would become an adaptation.) Traits evolving by random drift also belong to our first category. Second, there are traits that might be called epiphenomenal. These are traits that have evolved not on their own merit, as it were, but due to their connection to other evolving traits. The two simplest types of connections are gene linkage and pleiotropy. Third, there are traits that are present out of physical necessity. A trivial example, but one that well illustrates the type, is the fact that all organisms have mass. One need not and should not seek an adaptation-explanation of that fact. Investigation of less trivial examples is a current research program of considerable interest.[8]

I should point out that my explication of the notion of adaptation is in essential agreement with the influential analysis of G. C. Williams (1966). As far as I can tell, Dawkins would accept it as well. Burian (1983) is also largely in agreement.

So the proper objects of what-for questions in biology are adaptations, and adaptations are features of interactors that have evolved as the direct product of natural selection. What, then, is the structure of an adaptation-explanation? Basically an adaptation-explanation is a causal historical explanation that explains the presence and/or prevalence of an adaptation in terms of the selection forces leading to its evolution. An essential part of such an explanation is a citation of the effects of the adaptation that in fact increased the adaptedness of its possessors:

Put abstractly, a what-for question asked of adaptation A is answered by citing the effects of past instances of A (or precursors of A) and showing how these effects increased the adaptedness of A's possessors (or the possessors of A's precursors) and so led to the evolution of A (Brandon, 1981, p. 103).

Given the above analysis, do adaptation-explanations legitimize talk of adaptations as "for the good of something"? There is a sense in which they do. The evolution of an adaptation is explained in terms of the selective advantage its possessors (or possessors of its precursors) have over their competitors lacking the adaptation (or its precursors). That is, traits have effects on the adaptedness of their possessors; those traits that increase the adaptedness of their possessors and evolve because of that are adaptations. So, with selection as the arbiter of the good, we can say that adaptations are for the good of their possessors, and this locution serves an explanatory function.

But what sorts of entities are the possessors of adaptations? To think, for the moment, in terms of organismic selection, we might naturally think of organisms as the possessors of organismic adaptations. And so we would say generally that an interactor is the sort of entity benefited by adaptations. But couldn't we also say that the genes causally responsible for the ontogenetic production of the adaptation are the possessors of the organismic adaptation, and that they are ultimately the entities benefited by it? If so, then we would say, with Dawkins, that active germ-line replicators are the ultimate units of selection (what Dawkins calls optimons).

Can we resolve this question? We can, I think, if we properly understand natural selectionist explanations of differential reproduction. To explain evolution by natural selection one must explain its mechanism—the differential reproduction of heritable variation. But what is it that reproduces differentially? To fix ideas let us again consider organismic selection for increased height. In such a case taller-than-average parents tend to have greater reproductive success than shorter-than-average parents. So individual organisms reproduce differentially. But again, if height is genetically heritable, certain germ-line genes (the genes "for" increased height) have greater reproductive success than their alternative alleles. So genes reproduce differentially.

In the above case both interactors (organisms) and replicators (genes) reproduce differentially. How is this differential reproduction to be explained? The case in question is one of directional selection for increased height, so taller organisms are better adapted to this environment than are their shorter conspecifics (everything else being equal). A natural selectionist explanation of differential reproduction is one given in terms of the differential adaptedness of the reproducing entities to a common environment (Brandon, 1978 and 1981b). Should the differential reproduction of the organisms be explained in terms of the differential reproduction of the genes or vice versa? Or is this just a matter of taste and interest? I will argue that in this case, and in general, the differential reproduction of replicators is to be explained in terms of interactor selection.

In my argument I will employ the Salmon-Reichenbach notion of *screening off* (Salmon, 1971). By definition, A screens off B from E if and only if

$$P(E, A\cdot B) = P(E, A) \neq P(E, B)$$

(read "P(E, A·B)" as the probability of E given A and B). If A screens off B from E then in the presence of A, B is statistically irrelevant to E, i.e., $P(E, A) = P(E, A\cdot B)$. But note that this relation between A and

B is not symmetric. Given B, A is still statistically relevant to E, i.e., $P(E, B) \neq P(E, A \cdot B)$. Thus, where A and B are causally relevant to E, it follows that A's effect on the probability of E acts irrespective of the presence of B, but the same cannot be said of B. The effect B has on the probability of E depends on the presence or absence of A. For our purposes the important point is that proximate causes screen off remote causes from their effects.

Consider again our case of directional selection for increased height. As we have seen, in that case there is differential reproduction of interactors (organisms) and replicators (genes). Presumably everyone will admit that the mechanism by which genes replicate differentially in this case is the differential reproduction of organisms. So the fact to be explained is that taller organisms tend to leave more offspring than shorter organisms. Using the notion of screening off, we can easily see that this is best explained in terms of differences in height rather than differences in genotype.

Clearly, for any given level of reproductive success n, phenotype p and genotype g, $P(n, p \cdot g) = P(n, p) \neq P(n, g)$. This is, I think, a way of precisely stating Mayr's dictum that "natural selection favors (or discriminates against) phenotypes, not genes or genotypes" (Mayr, 1963, p. 184). The point is that an organism's having a certain height causally influences its reproductive success irrespective of its genotype. This is not to say that having a certain genotype is causally irrelevant to having a certain level of reproductive success; just that it is causally irrelevant given the organism's phenotype. Put metaphorically, natural selection sees a five-foot-tall plant as a five-foot-tall plant, not as a five-foot-tall plant with genotype g.

Just in case someone worries that our conclusion is a product of our choosing to look at differential reproduction of interactors rather than replicators, let me point out that the same conclusion holds even if we focus on differential reproduction of replicators. Still, in our example a particular germ-line gene's reproductive success is best explained in terms of the height of the organism in which it is housed, because again phenotype will screen off any intrinsic property of the gene (or the genotype) from its level of reproductive success.

What has been said about the above case is perfectly general. Interactors, by definition, are proximate causes, while active replicators are remote causes of differential reproduction. So any natural selectionist explanation will have to be given in terms of properties of interactors. Thus if the locution "Adaptation A is for the good of x" is to have any explanatory import, the values of x must always be interactors, the entities directly exposed to natural selection—the entities in terms of which natural selectionist explanations must be given. To use Dawkins's term, optimons will always be interactors.

All this is best appreciated by considering examples of selection occurring at levels other than that of the organismic phenotype. I have argued elsewhere (Brandon, 1982) that levels of selection can be characterized in terms of screening off. Selection occurs at a given level if and only if (1) there is differential reproduction among the entities at that level; and (2) the adaptedness values of these entities screen off the adaptedness values of entities at every other level from reproductive values at the given level. So, for instance, group selection occurs if and only if: (1) there is differential reproduction of groups; and (2) group adaptedness values screen off the adaptedness values of entities at any other level from group reproductive success.[9]

The importance of group selection as a force in nature is controversial. But the point I wish to make is conceptual. If group selection does occur, then group adaptations can evolve, and we certainly want to distinguish group adaptations from organismic adaptations. That is, we need to distinguish a group of adapted organisms from an adapted group of organisms (as is made clear by G. C. Williams, 1966).

The replicators in group and individual organismic selection do not differ. They are genes (Williams, 1966; Dawkins, 1976).[10] If we agree with Dawkins and say that adaptations are for the good of replicators, then we will be unable to distinguish group and individual adaptations, thus depriving such talk of all explanatory significance. That is, if adaptations are to be distinguished in terms of what they are good for, and if adaptations are for the good of replicators, then all biological adaptations would be essentially the same and we could not distinguish organismic from group adaptations.

Group and individual adaptations may look rather different. Much of the group selection controversy has centered on genetic altruism and how it could evolve by group selection. Genetic altruism could not, by definition, evolve by individual organismic selection. But the important distinction between group and individual adaptations is that they have different causal histories. The two processes could give rise to the same result.[11] For instance, although perhaps biologically implausible, it is possible that clutch size regulation in birds could arise by group selection. It is also possible that it could arise by individual selection (Lack, 1954). If we think such a trait is an adaptation, and if we want the locution This adaptation is for the good of to have any explanatory import, then we will have to characterize the adaptation as being for the good of interactors. The level of the interactors is determined by the actual causal process that produced the adaptation in question. It should not concern us that an organismic adaptation may fortuitously benefit a group, nor that a group adaptation may fortuitously benefit organisms; all that is relevant is the causal history of the trait.

What has been said about group selection is equally applicable to levels of selection where the interactors are less inclusive than individual organisms. Although the replicators may be the same, we again want to distinguish chromosomal adaptations, for example, one leading to meiotic drive, from organismic adaptations. In such a case the interactor level would be that of the chromosome. (Genes or collections of genes have physical properties—phenotypes, if you will—and so it shouldn't be surprising that they can function as interactors in some selection processes.)

I have argued that optimons are interactors, not active germ-line replicators. That is, when we say that interactors are for the good of something, that something is an interactor. My argument is based on an explication of adaptation-explanations that gives explanatory significance to talk of adaptations as being for the good of something.

I can think of only two ways of countering this argument. The first is to give an alternative account of adaptations in which talk of active germ-line replicators as optimons would have explanatory significance. I for one do not think that this can be done. The second is to deny that such talk is intended to have explanatory significance in the first place, and that it simply states a basic fact. But what is that basic fact? That adaptations aid replicators in reproducing? It is true that they do, but the same could be said of interactors; that is, adaptations also aid interactors in reproducing. (Recall the Hull-Dawkins disagreement over whether organisms could ever be considered replicators.) Is it that replicator reproduction is a copying process that is more direct and more accurate than interactor reproduction? That is true, but what this has to do with active germ-line replicators being the entities ultimately benefited by adaptations escapes me. In any case I find it hard to legitimize talk of adaptations as being for the good of something if that locution is to have no explanatory significance.

Conclusion

Why should anyone care about the nature of the things benefited by adaptations? The question, I would suggest, is fundamental to our basic understanding of nature. Prior to Darwin the appearance of design in nature was the basis of the most persuasive argument ever for the existence of God. With God as beneficent designer, all aspects of nature were seen as being for the good. Darwin, of course, offered an alternative explanation of apparent design. For the most part Darwin and neo-Darwinians have thought of adaptations as being produced solely by the process of natural selection acting at the level of individual organisms. For these Darwinians adaptations are for the good of the organisms possessing them, not for the good of groups, species, or ecosystems. Thus a Darwinian could not speak of bees making honey in order to provide food for bears. This is a radically different conception of nature from that of the creationist and it leads to a radically different perception of nature.

If we admit the possibility of selection occurring at levels other than that of the individual organism (and I don't see how anyone can at present deny that possibility), then we must admit the possibility of adaptations occurring at other levels. And, as I've argued, if we are to have an explanatory theory of adaptation, these adaptations at different levels must be carefully differentiated. The perception of nature that results from this hierarchical position is significantly different from that of the classic Darwinian. Recognizing the inevitable conflicts that will arise among various levels of selection, this view of nature is perhaps less adaptationist than the classic view. Certainly it is different in that it does not assume that every adaptation is for the good of the organisms possessing them.

Dawkins's view is different still. In a way it is similar to the classic Darwinian view. It too fails to distinguish among the effects of different levels of selection. But unlike the classic Darwinian view, it does not seem to be potentially explanatory of selection at any level.

The view that all adaptations are ultimately for the good of active germ-line replicators is strange. But in science the strangeness of a view can hardly count against it. In this chapter I have argued that the view is not useful, that there is a better conception of adaptations in nature.

Notes

I thank James Collins, David Hull, and James Lennox for helpful comments on an earlier version of this paper.

1. For example, see Ayala (1970), Wimsatt (1972), or Brandon (1981).

2. See Pittendrigh (1958), Mayr (1961), and Williams (1966).

3. Although Hull and Dawkins are for the most part in agreement on this distinction, there are some differences, including a terminological one. Dawkins uses the term *vehicle* rather than *interactor*. Their differences will be discussed below.

4. I say *may be* because cluttering up the genome with too much junk DNA could lower the probability of replication.

5. As Dawkins (1982a and 1982b), Hull (1980 and 1981), and Brandon (1982) have shown, there is not a unitary unit-of-selection question; the phrase unit-of-selection is ambiguous. One sense has to do with interactors, the other with replicators. I will argue that the sense relevant here is the one having to do with interactors. For more on this distinction see the introduction to Part II in Brandon and Burian, eds. (1984).

6. See the *Posterior Analytics* II 1–2 for a particularly clear statement of this approach. For his application of it to biological examples see the *Parts of Animals* II 14–16.

7. For example see Bromberger (1963 and 1966), Shapere (1974 and 1977), or van Frassen (1980).

8. For example see Kauffman (1983) and this volume.

9. I should point out that I consider this definition to be in complete agreement with what I take to be the dominant population genetic definition of groups, which is: "A group is the smallest collection of individuals within a population defined such that genotypic fitness calculated within each group is not a (frequency-dependent) function of the composition of any other group" (Uyenoyama and Feldman, 1980, p. 395). Thus what I will say about group selection will be equally applicable to interdemic and intrademic models of group selection. For reviews of these models of group selection see Wade (1978), Uyenoyama and Feldman (1980), Wilson (1983), or the introduction to Part III in Brandon and Burian, eds. (1984).

10. In this sense a gene is a hunk of the genome, the size of which depends on the amount of epistasis and gene linkage and the strength of selection. A gene in this sense could be the entire genome.

11. For experimental examples of this see Wade (1976, 1977, and 1979).

References

Ayala, F. J., 1970. Teleological explanations in evolutionary biology. *Phil. Sci.* 37: 1–15.

Brandon, R. N., 1978. Adaptation and evolutionary theory. *Stud. Hist. Phil. Sci.* 9: 181–206.

Brandon, R. N., 1981a. Biological teleology: questions and explanations. *Stud. Hist. Phil. Sci.* 12: 91–105.

Brandon, R. N., 1981b. A structural description of evolutionary theory. In *PSA 1980*, vol. 2, P. Asquith and R. Giere, eds. East Lansing, Michigan: Philosophy of Science Association, pp. 427–439.

Brandon, R. N., 1982. The levels of selection. In *PSA 1982*, vol. 1, P. Asquith and T. Nickles, eds. East Lansing, Michigan: Philosophy of Science Association, pp. 315–323.

Brandon, R. N., and R. Burian, eds., 1984. *Genes, Organisms, Populations: Controversies over the Units of Selection.* Cambridge, Massachusetts: The MIT Press. A Bradford book.

Bromberger, S., 1963. A theory about the theory of theory and about the theory of theories. In *Philosophy of Science: The Delaware Seminar*, vol. II, B. Baumrin, ed. New York: Interscience, 79–106.

Bromberger, S., 1966. Why questions. In *Mind and Cosmos: Explorations in the Philosophy of Science*, R. Colodny. ed. Pittsburgh: University of Pittsburgh Press, pp. 86–111.

Burian, R., 1983. Adaptation. In *Dimensions of Darwinism*, M. Grene, ed. Cambridge: Cambridge University Press, pp. 287–314.

Dawkins, R., 1976. *The Selfish Gene.* Oxford: Oxford University Press.

Dawkins, R., 1978. Replicator selection and the extended phenotype. *Zeitschrift für Tierpsychologie* 47: 61–76.

Dawkins, R., 1982a. Replicators and vehicles. In *Current Problems in Sociobiology*, King's College Sociobiology Group, ed. Cambridge: Cambridge University Press, pp. 45–64.

Dawkins, R., 1982b. *The Extended Phenotype.* Oxford: Freeman.

Doolittle, W. F., and C. Sapienza, 1980. Selfish genes, the phenotype paradigm and genome evolution. *Nature* 284:601–603.

Hempel, C. G., 1965. *Aspects of Scientific Explanation.* New York: The Free Press.

Hull, D., 1980. Individuality and selection. *Annual Review of Ecology and Systematics* 11: 311–332.

Hull, D., 1981. Units of evolution: a metaphysical essay. In *The Philosophy of Evolution*, U. L. Jensen and R. Harre, eds. Brighton: Harvester Press, pp. 23–44.

Kauffman, S. A., 1983. Filling some epistemological gaps: new patterns of inference in evolutionary theory. In *PSA 1982*, vol. 2, P. Asquith and T. Nickles, eds. East Lansing, Michigan: Philosophy of Science Association.

Lack, D., 1954. *The Natural Regulation of Animal Numbers*. Oxford: Clarendon Press.

Mayr, E., 1961. Cause and effect in biology. *Science* 134: 1501–1506.

Mayr, E., 1963. *Animal Species and Evolution*. Cambridge, Massachusetts: Harvard University Press.

Mayr, E. 1978. Evolution. *Scientific American* 239: 46–55.

Orgel, L. E., and F. H. C. Crick, 1980. Selfish DNA: the ultimate parasite. *Nature* 284: 604–607.

Pittendrigh, C. S., 1958. Adaptation, natural selection, and behavior. In *Behavior and Evolution*, A. Roe and G. G. Simpson, eds. New Haven: Yale University Press, pp. 390–416.

Salmon, W. C., 1971. *Statistical Explanation and Statistical Relevance*. Pittsburgh: Pittsburgh University Press.

Shapere, D., 1974. On the relations between compositional and evolutionary theories. In *Studies in the Philosophy of Biology*, F. Ayala and T. Dobzhansky, eds. London: Macmillan.

Shapere, D., 1977. Scientific theories and their domains. In *The Structure of Scientific Theories*, F. Suppe, ed. Urbana: University of Illinois Press, pp. 518–565.

Uyenoyama, M., and M. W. Feldman, 1980. Theories of kin and group selection: a population genetics perspective. *Theoretical Population Biology* 19: 87–123.

van Frassen, B. C., 1980. *The Scientific Image*. Oxford: Clarendon Press.

Wade, M. J., 1976. Group selection among laboratory populations of *Tribolium*. *Proceedings of the National Academy of Sciences* 73: 4604–4607.

Wade, M. J., 1977. An experimental study of group selection. *Evolution* 31: 134–153.

Wade, M. J., 1978. A critical review of the models of group selection. *Quarterly Review of Biology* 53: 101–114.

Wade, M. J., 1979. The primary characteristics of *Tribolium* populations group selected for increased and decreased population size. *Evolution* 33: 749–764.

Williams, G. C., 1966. *Adaptation and Natural Selection*. Princeton: Princeton University Press.

Wilson, D. S., 1983. The group selection controversy: history and current status. *Annual Review of Ecology and Systematics* 14: 159–187.

Wimsatt, W. C., 1972. Teleology and the logical structure of function statements. *Studies in History and Philosophy of Science* 3: 1–80.

Complexity and Closure

C. Dyke

An accumulating body of literature evidences the critical pressure now being applied to the orthodox neo-Darwinian synthesis. The pressure stems largely from three problem areas. A good deal of it comes from the attempt to integrate the new knowledge of molecular biology into an overall evolutionary picture. But attempts to build a comprehensive evolutionary ecology and to extend Darwinian explanation into the realm of social phenomena also contribute to the development of the critique. As far as I can see, none of these attempts denies the primary insight of the Darwinian tradition, the centrality of natural selection. Rather, each attempt resists two ancillary tendencies in the orthodox tradition: the tendency to reduce all explanation to a single level; and the tendency to totalize natural selection as the only operative explanatory principle.

One of the favorite explanatory rubrics of the orthodox evolutionists (especially favored by sociobiologists) is game theory. In what follows I will try to show how game theory must be used for it to be successful in explaining evolutionary processes, especially the sociobiological. This will require a three-pronged argument. First, some important methodological issues will have to be addressed; second, an adequate appreciation of the complexity of evolution will have to be developed; and third, the essential importance of closure and boundary conditions will have to be argued for. I conclude that only if embedded in a satisfactory methodology, with careful attention to hierarchical complexity, and with explicit statement of closure conditions is game theory a useful tool for the evolutionary theorist.

Problems of Method

Excellent historical research (Mayr, 1977; Ruse, 1971, 1975; Gillespie, 1979; Ospovat, 1981) has brought us to the point where we can now see clearly the overarching world view that formed the organizing background of Darwin's theory: the assumption of a largely indifferent universe within which inherited chance variations occur among organisms that compete with one another for scarce resources. We also know that this world view had its immediate source in Malthus. But Darwinism had then to be integrated into positivist science. For this to happen, the principle of natural selection, based on Malthusian competition, had to be detached from its origins and subjected to testing on its own.

As we look at the more than hundred-year history of formulations and reformulations of the principle of natural selection, we find, I think, that the evolutionary trajectory of Darwinian orthodoxy has been governed by the conflict between two sorts of commitment. The first is the commitment to a linear, bottom-up, derivation of the principle. This commitment is usually associated with empiricism, and with a concomitant belief in induction as the dominant reasoning pattern in science. The second is the commitment to a linear, top-down, justification of the principle. This strategy is usually associated with rationalism, and especially with various forms of idealism. The bottom sought in the bottom-up strategies is often referred to as raw data. The top referred to in top-down strategies varies somewhat from theory to theory. In the case of the principle of natural selection, the topmost principle is a general conception of nature as the arena for a competitive struggle between individual organisms. The adoption of this general Malthusian principle leads to the attempt to discover the particular competitive circumstances which account for evolutionary events.

Bottom-up strategies and top-down strategies persist together in the methodology of evolutionary biology because each strategy, taken by itself, is known to be inadequate.

On the basis of the positivist canons generally accepted, any set of raw data, no matter how large, is insufficient to establish a principle. Data is in general overdetermined in the sense that for any given body of data several or even many explanatory theories are consistent with it. As long as new data continue to fit the theory that we prefer, we can continue a research program based on that theory with some confidence. We haven't been proved wrong. But when, as at the present time in evolutionary theory, there are viable competing models, then, on positivist grounds, there is a standoff and we are free to pursue research on any of several different bases.

The further difficulty with the positivist bottom-up strategy has been pointed out over and over again in the literature, often in explicit relation to concrete research. The difficulty constitutes, in fact, a refutation of the possibility of a bottom-up strategy. A bottom-up strategy depends on the recognition of the bottom, the basic data identifiable independent of prior theoretical commitments. Otherwise the theoretical commitments constitute the top of a top-down strategy. But there is no such theory-independent bottom. Hence purely bottom-up strategies are impossible. The general impossibility of bottom-up strategies has been known since Hume, but the search for them has continued, since the difficulty of establishing top-down strategies has been apparent for at least as long.

If we are to derive a scientific principle or law such as the principle of natural selection from the top down, then we will have to choose a top, namely, some comprehensive principles. We are also going to have to defend our choice of principles. They cannot be plucked arbitrarily out of midair. When there are two or more plausible candidates for our choice, some way of deciding between them is imperative. No absolutely satisfactory way of doing so has ever been found, despite centuries of endeavor by rationalists. When, as is very often the case, all experimental data fit equally well (though not perfectly) with two or more tops then the problem of adjudication gets particularly acute, for we cannot, in that case, have recourse to bottom-up strategies to aid us. Thus is generated the problem of ideal types, well known in social theory, but seldom discussed with respect to theories in the natural sciences. Ideal types (Weber, 1947) are abstract descriptions of situations, phenomena, or persons that indicate the general features on which a theorist will focus as crucial for purposes of explanation. Obviously any situation, phenomenon, or person is of indefinitely many types. So the choice of one type as the relevant one involves a deep theoretical commitment.

Failure to talk about ideal types in the natural sciences is simply a matter of the history of terminology in the various disciplines. In the natural sciences the issues usually arise in terms of models (variously defined), especially alternative mathematical models. Nonetheless, it is worth importing Weber's term in order to emphasize the epistemological, ontological, and ideological features of many models and the groups of their acceptance. This is especially true of the Darwinian model, which has its origins in Mandeville's *Fable of the Bees* (Mandeville, 1723), travels through Malthus and Darwin to Wilson's marvelous work on hymenoptera (Wilson, 1971), and is then extended to human societies, thus closing the circle.

So long as competing ideal types are considered entirely from within the positivist framework, there is no definitive systematic way to decide between them. Each can be held onto tenaciously. Deviations from the results expected on the basis of the ideal type are handled by ceteris paribus clauses, more or less arbitrary adjustment of boundary conditions and the like.

The combined problem of the inadequacy of bottom-up strategies and the lack of definitive means of adjudicating among competing ideal types has been the major factor at work in the breakdown of positivist philosophy of science. In the face of this breakdown new strategies for dealing with methodological issues have emerged. Some of these strategies are genuinely new. Others are modifications of views conceived in the past. The most basic dimension of these strategies is a recognition of the diachronic aspects of investigative praxis. Scientific activity is seen to be one among many human activities. It has its practitioners, its norms, and its canons of success and failure—all of which are subject to modification over time.

The new strategies reject the positivist claim that theories are to be evaluated solely on the basis of atemporal criteria. What this means, in the end, is that the investigative process is itself treated as an investigatable subsystem with an internal dynamic, and with relationships to other investigated subsystems. The investigative process is both constrained by and constrains the objects of its investigations. From this point of view, methodology, epistemology, and ontology all are permanently susceptible to being put into question. Methodology I take to be the normative canons of research; epistemology the critical canons governing our choice of methodology; and ontology the set of objects of investigation presupposed by the methodology.

The expectable complaint from Platonists, at this juncture, is that I here threaten to historicize epistemology and ontology. This is not a complaint I have to take at all seriously. The only epistemologies and ontologies known are historico-cultural artifacts. Two-thousand-odd years of avowed and closet Platonist attempts to dehistoricize them have met with flat failure. Only theological canons of legitimacy demand dehistoricization. As long as those canons were hegemonic it looked as though the burden of proof were on the historicizers. No longer. Indeed, science as it is practiced at the bench and in the field has no necessary commitment to dehistoricized epistemology and ontology. Only the theology of science does.

Perceived from this historical point of view, methodology, epistemology, and ontology are three interpenetrating organizers of investigative activity. Our methodological, epistemological, and ontological commitments constrain our investigations and prevent them from being anarchic guesswork. This is so even when we concede that our current commitments are themselves subject to criticism and revision. You might say that our ability as a species to understand our world better and better depends on preventing our investigative activity from becoming a random walk. Our methodology, epistemology, and ontology, provisional though they be, have an essential role in bringing order to our attempts to learn and know.

To bring out this role I will introduce a terminology that may appear clumsy at the outset, but which will later serve to make essential connections. I will refer to methodology, epistemology, and ontology as "structured structuring structures" (Bourdieu, 1977). They are structured since they are the ordered consequences of prior rational investigative activity, and are inherited from this activity. They are structuring since they guide research and form the basis of further investigative strategies, constraining the range of what will count as further rational investigation, thus making possible and limiting what can be learned at any next stage. They are structures in an architectural sense which is barely metaphorical; like stud walls and rafters, they are determinate organizers of possibility.

When thought of as structured structuring structures, methodology, epistemology, and ontology become part of a thoroughly secularized dialectic in which none of the constituents of the investigative process is immune from critique in the context of a constant structured restructuring. In addition, every investigatable subsystem—physical, biological, cultural, social, etc.—is a potential contributor to the restructuring of future investigative access. For example, some structural features of the social system within which orthodox evolutionary theory arose are partially responsible for the acceptance and stability (hence both success and limitations) of the Darwinian research program. On the other hand, it is impossible to deny the effect that modern science has had on our conception of what there is in the world, or on our very conception of knowledge. Consequently we have to be careful of the way in which we locate Darwinism within broader social traditions. On one hand, the pride felt by Darwinians in their contribution to human understanding is well justified. On the other, we have to be sensitive to features of the Darwinian program consequent upon the presence of exogenous constraints.

We also have to be careful to have the patience to work with the constantly dynamic structured restructuring of structures. Structures are almost always in the process of building up or breaking down, or in some other way of being modified. Neither do they possess, in general, any essential identity-conferring features which necessarily persist through all their modifications. This means that if we can't trace their historical path we may not be able to work with them at all. One of the advantages biologists have in this regard is that they are already used to dealing with exactly analogous evolutionary transmogrifications.

Once we are able to look at science as an activity, scientific investigative practice is recognized to be, like all human practices, embedded in a social matrix. Thus science, conceived as a subsystem of social practice, is potentially subject to constraints imposed by the larger social whole of which it is a part. It is clear that this fact has to be resisted by science as a matter of policy. Indeed, Western science has, at least since Galileo, wanted to be free of any such social constraints. It has pursued a combination of strategies designed (sometimes consciously, sometimes not) to win and defend its autonomy both as an activity and as an institution. It is in this light that reductionism can best be understood. For reductionism is not merely a normative canon of science. It is also a strategy for sealing science off from possible external constraints. It does so in the first instance by insisting dogmatically that the phenomena with which science deals are themselves immune from social influence—eternal objects long antedating human presence. The social inputs upon the systems studied by science are, in other words, set at zero.

For this assumption to be plausible, science must claim to be a totalizing activity, to be the sole legitimate source of truth about the ultimate constitution of the universe. Abstraction and reduction are performed in order to make such totalization possible. In particular, physics has been extremely successful in establishing its program as the authoritative source of truth about the fundamental nature of the universe. This success provides the impetus for attempting to reduce all other science to an atomistic base, for physics could then pass its authority up the line to other branches of investigative activity.

Reductionism assumes that the progress of science is detachable from the process of science. That is, it assumes that claims to scientific truth can be certified within the internal canons of science itself, but that the resulting certification commands standing in intellectual life as a whole. This is the megalomaniacal claim to which otherwise cautious, modest, responsible scientists are forced when they are dragged out of their labs to participate in the dialectic of legitimation taking place in the intellectual community at large.

In this light, science's defense of its activity in the public forum can be looked at as an incidental activity only by accepting science's claim to autonomy a priori. Otherwise it must be considered an integral part of scientific activity seen as one among many activities practiced in and for a society which has the right of intelligent review. A society is ill-served by a science that seals itself off totally, just as science is ill-served by a society that won't allow it enough autonomy to pursue research.

We are fortunate to have a very recent example to consider. *McLean v. Arkansas Board of Education*, matching scientists against creationists, was most interesting in showing up the current dynamic of science and society.[1] Science has to defend its authority, its command of our belief, its command of a large share of our resources. It has to do so in a society under the sway of an ideal of religious truth, an ideal that retains its legitimacy for reasons science cannot touch. Give this background, intellectual authority is gained by those who successfully claim to be the source of truth. Scientists testified in *McLean* as authorities. That is, despite themselves, they testified as sources of truth. Yet there is the other, liberal democratic base that the sciences have to touch. The ideology of science demands that it be perceived as a community of open-minded equals engaged in a quest in which each success is provisional and revisable in light of successes at later stages of the quest.[2] But then any successes are vulnerable to the criticism of the absolutists. These critics must somehow be delegitimized. Somehow science must be made immune to all but criticism from within the scientific community itself.

So it turned out that scientists attempting to legitimize their authority while insisting on their fallibility proved to be a fascinating spectacle in Arkansas. In fact, the only way science managed to make ends meet (provisionally) was to get Judge Overton to buy into the doctrine of falsifiability, and a self-serving interpretation of how that doctrine ought to be applied. It was a lucky break. But the fault is not so much that of science itself, but rather that of the society which demands that its science compete for authority on the basis of religious criteria of legitimatization. We all complain of the religious excesses of scientific reductionism. We seldom notice the social and ideological pressures that give rise to them.

Science's conception of itself is that it has both won the authority of truth and risen above the social context of debate. Neither is true. That is a simple empirical fact. When this becomes obvious, as in *McLean*, the lofty strategies adopted by scientists become dangerously dysfunctional. They are not likely to get as lucky again as they were with Judge Overton.

This context and background make it particularly useful and illuminating to look at reductionist versions of sociobiology very carefully. Anyone who wishes to reduce human society and culture to biological determination has to argue that all structures constraining possible action—including the meaning horizons that organize collective action and individual aspiration—can be articulated entirely in biological terms. Presumably this means that the presence of these structures can be explained on the basis of the biological substrate, and that once in place the structures do not take up a necessary role in further explanations of new structures or of actions performed under the constraints of these structures. In an exactly parallel way, anyone wishing to reduce biological phenomena to physical determination must argue that all biological structures can be articulated entirely in physical terms. Presumably this means that the presence of these structures can be explained on the basis of the physical substrate, and that once in place, biological structures have no necessary role in further explanations of new structures or of activity occurring under the constraint of these structures.

At this juncture the physicalist reductionist has an interesting problem on his hands. He wants to hold that all phenomena are reducible to the behavior of the physical substrate. Does he want to hold that his own theorizing is a phenomenon on a par with all others, or does he want to hold that his own theorizing is special, and not subject to reduction? Of course, if he chooses the second option he is no longer a reductionist, so we can assume that he chooses the first, and his own theorizing is, like every other phenomenon, reducible to the behavior of the physical substrate (Feigl, 1958 and 1967; Margolis, 1983).

But all physical phenomena we know are undergoing change over time. The reductionist's account of the substrate to which all phenomena are to be reduced is itself, on his own account, a physical phenomenon, hence subject to change. Hence his account is neither eternal nor necessarily immutable. Whether it remains the same over a period of time depends upon its stability conditions within the successive environments it encounters. The a priori claim that a given account of the reductionist base is the final or definitive account is therefore untenable on reductionist grounds. An explanation of its eternal stability must be provided.

No such explanation has ever been provided, nor is it clear where one could be found. Furthermore, any such argument, if found, would prove the eternal stability not of a basic physical substrate, but of an extremely complex phenomenon: the belief that such-and-such is the ultimate physical substrate. Why should this complex phenomenon and not others be immune from the vicissitudes of time? If it is, then it would be difficult to argue that it is a phenomenon on a par with other standard physical phenomena.

Close relatives of this argument optimistically conclude that physicalist reduction is incoherent, hence impossible. This conclusion is unwarranted, however. All that follows from the argument is that on physicalist grounds reductionism is a historical phenomenon subject to the same sorts of investigation appropriate for all other phenomena. Unless the reductionist can predict the future course of reductionist theorizing, and the future results of scientific activity with certainty, indeterminacies and instabilities will be introduced into any nondialectical investigation of the nature of the substrate, that is, any investigation that does not simultaneously investigate the investigative process itself. These indeterminacies and instabilities can never be eradicated without denying the historicity of conceptualization; a fortiori they can never be eradicated without denying the reduction of thought to the physical substrate. If reductive physicalism is stated nondialectically, it denies itself.

In these days, as we see by reference to the *McLean* decision, all respectable scientists insist on the provisional character of science. But if the point is really taken seriously, then science has to be seen as the behavior of a complex system of knowers and knowns, investigators and the results of investigations. Remarks about results are necessarily also remarks about the investigations which produced those results. The history of modern science itself tells us that some results modify the way the process of investigation is conceptualized. At the same time, some reconceptualizations of investigative procedures modify the identification and interpretation of results. We are always required to freeze ourselves in some historical slice of the investigative process, but we are also obliged to unfreeze and consider the process itself.

Ultimate certainty falls in the light of our understanding of science as a human praxis, that is, as part of the potential subject matter of science itself.

At this point many people are overwhelmingly tempted to talk of relativism, since they note that the concept of scientific truth has become relative to place in a historical process. Such talk, however, betrays nostalgia for some extrahistorical vantage point, or for an autonomous philosophical enterprise having as its justifying task the search for such a vantage point. As Plato explicitly tells us, the search for such a vantage point is a death wish. In the absence of this death wish, the term *relativism* is unavailable for useful work.

Once we attune ourselves to the historical state of scientific activity we are ready to focus on explicit programs of investigation. At any moment in ongoing research, there will be a set of models available for application to the phenomena being studied. Some of these models will have enough legitimacy to be live options. None of them will have enough legitimacy to be the only option. We know that we must pursue strategies that are consistent with our earlier successful practice and are legitimized by this success; take full advantage of the resources of the array of models we have at our disposal; don't foreclose any legitimate options; and don't leave us with a tangled mess of hypotheses incapable of being integrated or even compared.

Thus we have a problem of research design much more complicated than the hypothesis-testing paradigm familiar in the positivistic socialization of the young scientist. In fact, research design ceases to be an a priori activity aimed at herding a pre-existing array of bare phenomena. Instead, it becomes a coordinate and interpenetrating part of the investigatory problematic. In other words, the activity of investigation has to be considered as a whole, and any decisive delineation of investigator and object of investigation has to be won on the basis of the establishment of boundaries which have been argued for, not arbitrarily set out beforehand. In general, the establishment of investigator/object boundaries will partially be a function of the models chosen as guides to research. The initial attempts at establishing the boundary conditions will also be constrained, of course, by the success of previous research strategies.

It turns out that there are no general solutions to the research design problem. As I said before, the progress of research depends on the process of research. This means that we have to look at first-order research itself in order to have a starting point for the critical evaluation of methodology; and that critical evaluation must at least advance first-order research if it is to justify itself. I am sometimes told that this line of thought implies that there is no such thing as philosophy as an autonomous activity or discipline. I suppose that this is true. It is also trivial.

Closure

The general form of a closure condition is shown by the following:[3] Assume that A and B are the alternatives. Either A or B; not A; therefore B. The closure condition is the assumption that A and B are the only alternatives. While certainly not always as simple as this formulation suggests, closure conditions pervade our explanatory activity, and are often so well understood that they go without saying. When we do want to make them explicit, we can often do so rather easily. We say what we're talking about, that is, we specify a domain of discourse in a way that pins matters down in an acceptable way. Or, if we are engaging in formal experiments, we design them with an appropriate system of controls that serve to establish closure. Or, when we have well-established and well-developed scientific theories at our disposal, we can often derive closure conditions from them. Often boundary conditions yielding closure are consequences of the properties of particular mathematical formulations. A major aspect of the Galileo-Descartes-Newton achievement in the establishment of a mathematical physics consisted in identifying a small closed set of physical parameters sufficient to provide explanations. The solar system remains the outstanding example of a natural physical system that is closed with respect to its mechanics (though it clearly isn't closed from, say, an information-theoretical point of view). In some areas, such as orthodox thermodynamics, the specification of boundary conditions and other closure conditions is a constant task. The standard physical symmetries (conservation laws) require precisely defined closure.

Science, with physics as its de facto model, and experiment as its primary investigative technique, has made undeniable successes. But the adoption of physics as exemplary has meant that problems of systems closure have been treated in a relatively standard way. From a mathematical point of view, systems closure has been assumed to hold for a set of phenomena when a mathematical model of the phenomena could be produced that was formally similar to a model holding for a closed physical system. When a genuine question of closure arises, raised in terms of the adequacy of the model, an independent argument must be found to establish closure. But usually, closure is posited as a satisfactory approximation, ceteris paribus. This latter tactic is nearly universal in the construction of econometric models, but it also occurs frequently in population biology. The tactic is part of what motivates my insistence on thinking of such models as ideal types.

Closure conditions quite often appear to be *dimensional,* as when it is said that the solar system is mechanically closed and thermodynamically open. This dimensionality could disappear only if all the dimensions could be reduced to one. The Laplacian program was, among other things, an attempt to perform just such a reduction in Newtonian terms. Philosophers are familiar with problems of dimensional closure in issue after issue generated by the Galileo-Descartes-Newton program: primary and secondary qualities; the mind/body problem; the relationship between reasons and causes in human action; the establishment of identity *simpliciter* as opposed to identity under a given description. I think these are all problems involving closure and boundary conditions in complex hierarchical interactive systems, and are best studied as such. Here they serve as reference point analogues.

Questions of closure arise in evolutionary biology in ways exactly parallel to the issues just mentioned. In its progress over the past century, biology has dealt with some these questions with amazing success. As in physics, some problems of explanatory closure have lent themselves to solutions in terms of experimental design or in terms of theories specifying the range of operative variables. However, robust explanations within evolutionary biology have proved more difficult to obtain than was hoped. In my view, a good part the reason difficulties have arisen is that the complexity of evolutionary (and ecological) systems has not always been confronted with sufficient respect. In attempting to achieve explanatory closure in terms of simple, one-dimensional mechanical systems, orthodox neo-Darwinians have cut off access to a full understanding of the phenomena they study. This self-limitation was indeed for a long time necessary and productive, but as the program has tried to extend itself to issues in molecular evolution and sociobiology, the limitations have begun to look particularly serious.

The examination of game-theoretical models will serve to exhibit some of the worries I have about explanatory closure in evolutionary explanations. I think that there are interesting uses for game theory within evolutionary biology, but there are also major pitfalls.

Bookkeeping, Teleology, and Players

It is important to note that game theory is, in its standard applications, a bookkeeping theory and not, by itself, an explanatory theory. For instance, suppose you watch a game of tic-tac-toe and see both players pursue an optimal strategy. The game ends in a draw. You ask, why did the game end in a draw? You get the answer: Because both players pursued optimal strategy. This may or may not be an explanation, depending on what you have packed into *pursued*. There are innumerable reasons why someone playing tic-tac-toe might execute a sequence of moves that constituted optimal strategy from a bookkeeping point of view. (Learners sometimes do it by accident.) However, in this stupefyingly simple example, there is hardly any room at all for alternative explanations. The natural immediate inference from bookkeeping to explanation is virtually forced upon us, since the availability of plausible alternative explanations is virtually nil. However, the easy inference is an artifact of the trivial example, and especially of the fact that it is an example of a sequence of action that is utterly meaningless if not performed in a game, as a game, and for the sake of winning the game. The general meaninglessness of the activity provides closure yielding the inference to the one explanation that survives as a plausible candidate. More complicated activities must be treated with more sophistication. But in each case, any inference from game-theoretical bookkeeping to explanation requires the imposition of closure conditions.

For games in the literal sense, the usual conditions producing explanatory closure are that the players are playing to win; are under no external constraints; and are rational in the appropriate sense. These closure conditions added to the bookkeeping yield an explanation— though, as is often pointed out, they tend to yield tautology if care is not taken to leave open the possibility that, say, the players could have chosen not to play to win.[4]

A second sort of closure required in order to extract explanations from game-theoretical bookkeeping is what we could call payoff closure. We know that if we are to model a phenomenon as a formal game, we must be able to construct (or at least provide a recursive recipe for) a payoff matrix, and so we must also be able to make sense out of the entries in the matrix. For example, one of the most important measures used in sociobiology is Hamilton's notion of inclusive fitness (Hamilton, 1964). This measure defines relative success and failure in the evolutionary game in terms of numbers of descendant gene bearers. If this measure is accepted, then it appears that there is at least a hope of specifying payoffs for useful game-theoretical models. But there is serious doubt about whether this measure should be accepted. It follows immediately from its simplistic use that *E. coli* are more successful than human beings. A more sophisticated use would confine inclusive fitness to intraspecific comparisons, whereupon Catholics win the evolutionary game when they play against Protestants. We could, by brute force of will, decide that this was what evolutionary success meant—adopting thereby the rather odd teleological criterion of numerousness. We have to note, however, that bacteria, at least, have achieved their success while sacrificing complexity (and quite possibly by avoiding complexity). This presents us with an interesting choice, for some people wish to construct their evolutionary teleology in terms of higher and lower levels of complexity, with humans at the pinnacle of evolutionary success, despite the fact that they are less numerous than, say, earthworms.

The adoption of an evolutionary teleology of some sort will provide closure of the payoff matrix. Indeed, game theory works for games in the literal sense because the teleology of winning can normally be assumed. For evolutionary phenomena, assuming the teleology of numerousness, simply reflects a decision to have the entries in the game matrix represent numbers of offspring. If, in search of more sophistication, you produce a more complicated formula for constructing the matrix, taking into account such things as carrying capacity, this won't change the nature of the teleological closure condition. Presupposing the teleology of complexity, on the other hand, will require finding a measure of complexity and filling in the matrix on its basis. Several quite persuasive and perhaps useful measures of complexity have already been devised, e.g., information density (Gatlin, 1972).

There are several reasons for being careful about the difference between intraspecific and interspecific comparisons. Darwin's own theory was concerned almost entirely with intraspecific comparisons. The variation he presupposed was variation within a single species, and differential survival is almost always differential survival of members of the same species. Initially, then, we have to remember that speciation events occur. Darwin's focus—determined, no doubt, by his concern with the species problem as it was historically posed—is perfectly understandable. However, it has to be made clear that the selection of that focus functions effectively as a closure assumption, sealing off, as it does, interspecific interaction as a secondary consideration. This closure was extremely unstable. We only have to think of how quickly mimicry obtruded itself as a subject for Darwinian explanation to see that it was (Turner, in Grene, 1983). There may still be problems that yield to examination of solely intraspecific comparisons, but certainly no genuinely ecological problem does. Insofar as evolutionary ecology has shifted to the core of evolutionary biology, the original Darwinian program has had to be modified. I contend that the shift in models occasioned by the problem shift can profitably be looked at in terms of the various closure conditions required in order to extract explanations from the various models.

The concepts of fitness and adaptation are important here. A great deal has been written about them lately. I will fit them simply into the present discussion. All fitness measures, including inclusive fitness, behave formally like budget possibility surfaces in economics. Differences in fitness are exactly analogous to differences in budget. Just as any set of resources in economics can constitute budgetary assets only within a particular set of economic arrangements, exchange system, or the like, so fitness represents assets as defined in a particular environment. Just as in economics nothing, not even money, is a system-independent asset, so no trait is necessarily a contributor to fitness, especially at the margin. However, of course, in both spheres some things are more centrally and surely assets than others. They contribute to success in virtually every economic or ecological environment.

Closure conditions enter the consideration of fitness in expectable ways. Fitness is always fitness with respect to an environment or range of environments. Variation is evolutionarily relevant only when it contributes to the fitness budget relative to predetermined environmental variables. In principle there is no problem in specifying these conditions in the abstract. The trouble comes in matching the chosen model and the closure conditions selected to the situation being modeled. In other words, the number of possible evolutionarily relevant traits and relevant environmental conditions is very large. It takes a great deal of analysis even to get to the point of choosing an appropriate model. Any analysis in terms of fitness must deal with complexity and closure with great care, whether or not the model used is a game-theoretical model.

The assumption that each organism is either in an adaptive equilibrium with its environment or in pursuing an evolutionary path toward adaptation is one of the most common devices for closing ecological systems for purposes of explanation. Recent literature has amply shown, however, how easy it is for the concept of adaptation to fall into tautology or vacuous teleology (Brandon, 1978 and this book; Burian, 1983; Mills and Beatty, 1979). Most evolutionary biologists, of course, talk as if they would like to avoid teleological explanatory closure. Historically, post-Darwinian biology has prided itself on eliminating teleology from evolutionary theory. So it is surprising that subdisciplines such as sociobiology, professing to be Darwinian in spirit, rely on closure conditions such as inclusive fitness that are openly teleological. The best arguments against doing so are their own. The main one is that there is no convincing scientific way to establish one end rather than another as the correct or real one. At any rate, numerousness is hardly one of the attractive candidates if one is seeking a *telos*. In fact, it is so unlikely a *telos* that it fosters the illusion that it is not a *telos* at all.

Is there a nonteleological way to achieve payoff closure so that we will find game theory useful? We might try the one proposed by Slobodkin and others (Slobodkin, 1968, 1961/1980; Tuomi and Haukoija, 1979; Tuomi et al., 1983). Slobodkin speaks of the "existential game." The idea is that the only payoffs are survival and failure to survive. Survival needn't be thought of teleologically any more than the persistence or transience of any other physical object. Some water is ice, and some is not. Some will be ice for a long time because it is bound up in an arctic glacier; some will be ice only until the sun beats down on the windshield of my car. Neither quantity of water is a particularly successful piece of ice, nor is either to be disparaged for its failure. The persistence of a crystal and the persistence of an organism can be thought of in the same nonteleological terms. The dinosaurs were not evolutionary failures. There just aren't any around any more.

This secularized version of assigning game payoffs is promising in a number of ways. First of all, it relates easily to the basic sort of negative selection view favored by Darwinians. Second, it allows us to think of the principle of natural selection as the idiosyncratic articulation of one among many concatenations of stability conditions for natural objects at a certain level of complexity. Comparable stability analyses can be provided for atomic nuclei, crystals, chemical compounds, and so on up through a rich array of structures, including social structures. The principle of natural selection stands out, and appears qualitatively different only as a historical consequence of its genesis and of our special fascination with organisms, including ourselves. Finally, the existential game gives us a sensible way to deal with numerousness, complexity, longevity, and other important features of evolutionary phenomena. We can say that they constitute various strategies for achieving payoffs—or, if we are really serious about eliminating the teleology—we can say that they are explanatorily related to the relative persistence of various species, and go on to spell out the explanations.

The basic facts of the evolutionary game are that there are some organisms around; some that were once around but are no longer around; and (more than likely) some that have not been around yet, but will be around later. The matters to be explained by evolutionary theory are: Why are some organisms around and not others, and how do new organisms arrive on the scene? From the point of view of the existential game, the question is whether we can hope to explain who the survivors are likely to be, and why they are likely to survive.

Choosing the Players

If we are going to talk about who survives to play on, we will have to consider the players in the evolutionary game. The most likely candidates are individual organisms. Individual organisms live and die; and breeding takes place among individual organisms. So, classically, individual organisms are thought to be the players. There are, however, other candidates. Dawkins puts forward DNA as the player(s) (Dawkins, 1976). From a different perspective, Lila Gatlin (1972) also suggests the DNA molecules are the players, though her theory differs in crucial respects from that of Dawkins. Others have proposed groups, species, clades, organized independent entities such as hills of ants, etc. Eventually I want to say: All of the above plus others. But that point has to be arrived at slowly through another examination of closure conditions. We need to get closure on a statement to the effect that S, Y, Z, et al. are the evolutionary players, and they are the only players.

When game-theoretical explanations stand under teleological closure, the identification of players is quite naturally achieved by locating the teleological agents: The players are those entities who make moves in order to win. The trouble is, though, that with the possible exception of some hominids, none of the potential players in the existential evolutionary game cares whether it wins or not. In fact, precious few of the potential players have the slightest idea that they are even playing a game. Pretending to the contrary does make the use of game models much easier than otherwise (by introducing teleological closure), but at the same time it vitiates the attempt. The fairy-tale parallel to a theory need not shed much light on the quality of the theory itself.

This is yet another reason for rejecting teleological closure. So closure on the list of players has to be found elsewhere.

First, let us imagine a typical ecological complex and think about the things we could count. Unless some reason can be found for ruling them out (that is, an explicit closure condition), each will remain as a candidate for the list of players.

(a) There are species that can be counted—within limits. If speciation events are taking place within the complex, indeterminacies will arise. But the fact is that counting species (on the basis of identified individuals, of course) is one of the first steps in any census.

(b) Higher taxa can also be counted. There may not be very frequent occasion to consider them as players, but investigations of macroevolution might require that we model, say, the competition between angiosperms and gymnosperms.

(c) It is ambiguous whether the population of each species at one time is the same as the population at an earlier or later time. Population is a term too flexible to do its job without further specification. Do reproduction and death constitute a function that maps one population onto another population quasi-continuously? Or do they constitute a function that tracks changes in the same population? The term is used in both ways. Suitable identity criteria can be established to justify either usage. The precise audit in any given case awaits the adoption of one or the other of the sets of identity conditions. If, for example, we want to say that the population of a certain species has increased, we will need to adopt one set of identity conditions. If we want to say that a parent population has managed to give rise to a filial population, we will need the other. Many people seem to want to say both at the same time.

(d) Individual organisms can be counted. Our ability to do so depends once more on some very sophisticated theoretical judgments resulting in identity conditions. For example, any stand of trees in which all the individual trunks have grown from one continuous rootstock presents an interesting problem. In addition, our ability to count individuals of a given kind depends upon the theoretical taxonomic assumptions establishing the kind. Counting the dinosaurs that survive in the present day may require a joint effort involving (at least) paleontologists, evolutionary geneticists, and, depending upon the decision reached by those two groups, the Audubon Society and Frank Perdue.

(e) Breeding pairs can be counted. In sexually reproducing species, especially in small populations, this can be an exceptionally important account to keep. But then in species whose populations exhibit social structure there may well be crucial accounting units (herds, flocks, schools, etc.) of varying sizes that, for example, determine breeding statistics. Do individual antelopes play games against predators, or do the herds play team games?

We now move to a list of potential players least likely to be candidates for teleological closure.

(f) DNA, whether selfish or not, has been cast as a player in an evolutionary game. The problem is to etablish identity conditions for it that will yield a useful accounting. The situation is intriguing from a philosophical point of view because a consideration of DNA in this context plunges us immediately into a consideration of types and tokens. I think that right at the outset it is possible to reject individual pieces (tokens) of DNA as the players in any sort of game. This is especially true when molecular games are spelled out in thermodynamic- or information-theoretical terms. For the information differentials crucial to differential survival are at least as likely to be embodied in stochastic processes as deterministic ones. Coupled with the fact that individual tokens of any DNA type are different as the cells in which they occur are differentiated, this means that DNA eventually acts evolutionarily only through an internally structured complex, making the individual pieces of DNA more like the pieces in the chess game than the players. If it is remarked that the same can be said of individual organisms vis-à-vis breeding populations, I can but nod my head and note that both situations need to be worked out clearly in a more ample context. Here I will say only that a consideration of the difference between DNA as type and DNA as token is quite crucial for getting a firm grip on the difference between bookkeeping and explanation. It is clear that over the past twenty years, as people have taken the concept of a genetic language seriously, they have uncovered problems about the nature of primary information bearers that are identical to problems faced in more traditional philosophies of language (Margolis, 1983).

(g) The immediately foregoing could lead to polynucleotide/protein complexes as possible players in an evolutionary game. For it is known that such molecular complexes can bear information and engage in causal processes that naked DNA cannot (Kaufmann and Campbell, this book).

Let me stop the list of professional players here, though I am far from certain that I have included every important candidate. We have more than enough complexity in view already as we examine the questions of explanatory closure that concern us.

We now have to emphasize the system-dependence of the identity conditions for players as players, and a given statistical array as a payoff matrix of a game played by those players. They must be identified together. Atomists will quickly point out that the players, at least, are independently identifiable since, for one thing, we were able to identify individuals and populations long before we started thinking in terms of evolutionary games. This misses the point. First an example. There are species of insects of which the males and the females were long classified in two different species (and even in two different higher taxa). Caught in flagrante, biologists blushingly reclassified them. Similar stories are available with respect to juvenile and adult forms of the same insect. An identification is an answer to the question, What is it? Every identification is system-dependent. In the case of the insects just cited, breeding interactions and life cycles had to be known in order to arrive at satisfactory identification, even taxonomically. The system-dependence of identification is often ignored or denied since the sloppy, complex megasystem that grows up within everyday life doesn't seem systematic enough—or is so polysystematic that it seems to be asystematic. What is it? usually gets answered within the context of a way of life so familiar that the identification of things seems to be a recognition of what they are in some absolute system-independent sense. Yet any identification is an assignment of place in a (more or less coherent) order of things (Foucault, 1970). It is a historical artifact. There is no way that identity can be detached from identification except in ethereal philosophical abstractions. Every identification has its theoretical underpinnings in the evolving epistemological and ontological nexus of life—a life with practical, usually theological, and possibly scientific dimensions.

The debate within analytic philosophy about identity under a given description versus identity *simpliciter* scratches at the surface of the system-bound praxis of identification. Every argument (since Parmenides) to the effect that identity can be established *simpliciter* has been an argument that such identity must be established if a totalizing absolute system is to be possible. But this is a theological dream. An alternative project is to use the identification and re-identification of certain common things to investigate the evolutionary dynamics of our ability to integrate disparate systematic segments of our experience as our investigations proceed. This would include experience we make possible by means of the artifacts of our science: electron microscopes, radio telescopes, etc. The battle for primary identification within such a project would be the battle for legitimacy of one systematic praxis over others as the organizing praxis of our lives.

The best example for our present consideration is, again, the war between the creationists and the evolutionists. Each party must legitimize its own primary identification of, say, human beings, and delegitimize the primary identification put forward by the opponent. If humans are primarily the descendants of Adam, then this will impose fundamental constraints at the interface between theology and science. Furthermore, the constraints will penetrate into science, dictating the legitimate range of interpretation of experimental results and paleontological discoveries. If, on the other hand, human beings are fundamentally upshots of a long evolutionary process, then this will dictate a new range of hermeneutic possibilities available for dealing with the classic religious texts. As long as both identifications exist as fundamental organizers among alternative subcultures, the interpenetration is inevitable. There is no supertheory for adjudicating the differences. Nor is there any compelling reason for thinking that there ever will be.

This impinges on our discussion of game theory in the following way: To say that players and the game must be identified together is to say that the systematic constraints of the game model establish a local ontology specifiable only within the full account of the game. For some games the entities that turn out to be the players will already have been identified in some systematic way. This is no accident (it is a tribute to the ongoing success and expansion of our investigative praxis), but it is also no necessity (new things—e.g., galaxies, codons, quarks—may be discovered in the course of our investigations). Our discovery of game models may lead to the discovery of players. It would be sheer dogma to think that the entities involved in biological evolution (that is, the units of selection) must be precisely those biological entities previously fixed upon as the primary biological units. But, on the other side, the continuity of investigative praxis requires integration of any new projected units into the solid core of an already highly successful biological research program. New units cannot be fudged arbitrarily and have any legitimate claim to be taken seriously by working biologists. Burian's discussion of the gene provides a good illustration of this point (Burian, this book). This is why the list of potential players

I provide is confined to what I can find to be strong candidates on the basis of the literature of contemporary biology itself.

Thus the complexity of the situation makes straightforward identif-iction of games and players more difficult than the literature would suggest. We can see this most easily in a brief analogy. Baseball players are not, from a game-theoretical point of view, the players in the game of baseball, where the payoffs are pennants and world championships. The players are Cardinals, Orioles, Blue Jays, Tigers, Cubs, Angels and so on—that is, teams. The ballplayers themselves have a special claim to be thought of as players, because they are the ones who hit, throw, and field. Just so, individual organisms have a special place in the evolutionary game, because they are the ones who are born, forage, graze, prey, procreate, and die. But the ballplayers are embedded in a structured structuring structure that radically constrains their activity as players, dictates their roles in the game being played, and prevents them from being players of that game in the technical sense. The same is true of individual organisms.

From another perspective, ballplayers are sent onto the field or kept on the bench, given orders as to where to stand, whether to swing the bat, when to run, etc. The manager thus manipulates them, so from this perspective the ballplayers are more like the pieces in chess than they are like chess players. Like the chess pieces their roles are dif-ferentiated, occasioning an internal structure for the composition of the team. A manager can't win with twenty-five pitchers on the roster, for example. Or if the manager has a fast centerfielder, he can get away with a slow-footed rock-handed slugger in left field; otherwise not. Thus the selection of ballplayers for a team is seriously constrained by the structure of the game, and, in the last analysis, by the abilities and specialties of other ballplayers on the team and ballplayers on competing teams.

Every ecological environment is multistructured too. So any analysis of it in game-theoretical terms has to rest on a correct account of the structures. In fact, the baseball analogy is much too simplistic to model natural environments. A closer analogy would be one in which the manager tries to put together a roster to compete in baseball, basketball, football, and soccer in the same season.

The Games Evolution Plays

There is no reason to believe that a one-dimensional analysis focusing on individual organisms as the sole players will succeed. Such an analysis is a highly abstracted ideal type. Pursuing it demands imposition of an a priori structure that only fragmentarily reflects the goings-on in the natural environment. The proliferation of ceteris paribus clauses that occurs when the defenders of this ideal type are confronted with structural complexity is the tip-off to the inadequacy of the ideal type. In the end, adoption of the ideal type predetermines an arbitrary choice of closure and boundary conditions that impedes research rather than furthering it.

It is easy to see, though, why the ideal type of the individualist competitive game persists so strongly. It seems to offer a clean set of mathematical techniques to deal with the competition assumed to be the main organizing fact of nature. But the mathematics of game theory is only incidentally connected with competition. It merely organizes the differential outcomes of joint activities in matrix form, and examines the consequences of activities that would occasion different paths through the matrix. These activities could be competitive, cooperative, or neither. For instance, in the classic statement of the Prisoners' Dilemma, the players are not initially competitors, but, rather, self-interested and either indifferent to each other or slightly sympathetic. The district attorney changes the structure of their relationship. The Prisoners' Dilemma is instructive because it shows how structure can be imposed on two people so that they become, from a bookkeeping point of view, competitors. And, if there is an external agency powerful enough to keep that structure in place, the two players may well have to come to recognize each other as competitors and act accordingly, thus converting bookkeeping dialectically to explanation.

Obviously, climatic changes, migration, and most other ecologically relevant events can have the same effect. I see no reason why emergent social structures cannot do the same. On the other hand, evolutionarily important events could have the consequence of breaking the competitive relationship between players, rendering them functionally oblivious to one another, or even landing them in positive mutuality. If, when, and where this occurs is clearly a matter for concrete research, not a priori theorizing. Any use of game theory will require analyses at least sophisticated enough to bring all these structures and structured structurings to light.

Sometimes a relatively isolated, relatively self-imposed game will be found, especially in appropriate time frames in relatively stable environments. The search for such isolation and independence is, after all, very like the search for system integrity within a hierarchy of near-decomposable systems (Simon, 1981). In the usual way, horizontal and vertical uncoupling may exist so as to leave a particular subsystem relatively autonomous for a significant length of time. So, for example, a particular predator/prey interaction may end up being, for all intents and purposes, the subsystem of an ecological complex of systems, decisive for understanding the fate of the members of the system, especially the predators and prey themselves. The genomes of the two may be stable (the rate of change tiny compared to the rate at which they affect each other in the predation relation), the climate constant, the primary food source relatively constant, etc.[5]

This sort of simple closure is exemplified by Darwin's own treatment of genetic variability. His theory posits inheritable, randomly generated variability. This is to say, Look for no explanations of evolutionary events within the genome; attribute them to chance. Randomness is posited in order to achieve explanatory closure with respect to high-frequency events, and, at the other end, uniformitarianism is posited to achieve closure with respect to low-frequency events. In short, Darwin's theory assumes that middle-frequency evolutionary events are sealed off vertically. This allows him to achieve explanatory closure at the phenotypic level.

The language of "high, medium, and low frequency" is another borrowing from Simon (1981). The idea is that in any hierarchical system of systems there are events occurring at high frequency in the lower level, systems which can be lumped together, averaged, or otherwise summarized in a single parameter (or "sufficient parameter": Levins, 1968) for modeling the middle-level system. Futhermore, there are events occurring at low frequencies in the higher-level systems that are infrequent enough to be treated as occasional exogenous perturbations of the middle-level system. The concept is obviously derived from the investigation of physical systems, but has persuasive familiar analogues. In astronomy the mechanics of the solar system can be described without referring to events at the atomic level (high-frequency events) and without referring to events of interstellar evolution (low-frequency events). In meteorology, weather systems can be described without reference to the motion of air molecules, and without references to secular climatic changes. The analogous picture relating to Darwin involves high-frequency events at the molecular and chromosome level, and low-frequency events at the geological and climatological levels.

Clearly, the subsequent synthesis with Mendelian (and Hardy-Weinberg) statistics depends on the same sort of sealing-off assumption.

Mendel's laws of segregation and independent assortment are, in fact, closure conditions. They can be questioned quite reasonably, as they have been by Beatty (1980), who points out that segregation and assortment are the consequences of meiotic events that themselves may be (in fact must be) under genetic control.

Consequently, molecular and chromosomal events affecting linkage, crossing over, and other obstacles to Mendelian ratios cannot be sealed off in any simple way in order to allow explanation at one level only. Thus the move from the Mendelian bookkeeping to explanation is far less straightforward than the standard synthesis would have it.

There is no doubt that the early orthodox Darwinians were right in imposing middle-frequency closure. Ignorance of molecular genetics demanded it. Within its limits the program was enormously successful. The anomalous deviations from the standard ratios had to be allowed to pile up as the orthodox program was pushed as far as it could go. However, the orthodox theory is no longer faced with a heap of unsystematic anomalies. Molecular genetics, immunology, information theory, and other fields have cut away much of the ignorance that justified middle-range closure vis-à-vis high-frequency events. To a lesser extent, closure vis-à-low-frequency events has been challenged now that the uniformitarian hypothesis has a reasonable competitor in the theory of punctuated equilibrium. So the old closure assumptions now appear simplistic and arbitrary.

From another angle, the simplicity and arbitrariness of the orthodox Darwinian program is most glaring when the interpenetration of biology and society begins to be examined seriously. Since the neo-Darwinian pattern requires the confinement of explanations to a single systems level, and is geared up to handle only one-dimensional selection processes, it has no comfortable way to deal with the emergence of new social and cultural systems with an evolutionary dynamic of their own.

Reproduction is a biological phenomenon even when it occurs among humans; and differential rates of reproduction continue to constitute a biological measure in part. Nonetheless, a one-dimensional evolutionary explanation of the differential reproductive rates of the Roman Catholic laity and the Roman Catholic clergy would be quite bizarre. We need not rest our case on such examples, however. There are many reasons for resisting the reduction of social systems to biological systems. The reason that ought to have weight for biologists is that the standard reductions within evolutionary biology itself are unsuccessful.

Prospects for Closure

At this point it is possible to make the search for explanatory closure look like a hopeless one by harping on the potentially endless complexities of any investigated system of systems. I do not want matters to appear hopeless, though; I just want complexity to be confronted honestly. We can start at the level of epistemology. The short way to formulate the classic problem of epistemology as it emerges in the present context is: Can closure be achieved in the search for explanatory closure? The short answer is no. The Platonic/Cartesian dream of ultimate closure, especially in its Christian and Positivist forms, must be rejected. The romanticism of the hermeneutic circle and the Feyerabendian anarchism (Feyerabend, 1975) had better be rejected too. Peremptorily, I take it as given that we do get a better and better understanding of at least some of the phenomena we study. One can get to be a better scientist just as one can get to be a better tennis player, lover, or potter; and, with the exception of a possible Alcibiadean moment in one of these, none of them requires anything Platonic to account for improvement. Scientific progress is like biological evolution itself in this regard. That they occur is a fact. How they occur is a complex issue.

Scientists themselves (as we've seen in *McLean*) insist upon the provisional, ongoing, improving character of their activity. So it should be reasonably comfortable for them to consider that the simplifications and abstractions that guide their activity profitably during one period of scientific advance can be put into question at a later stage. Another way to put one facet of this is to say that the search for explanatory closure is never closed. This is not, however, disabling. At any given time reasonable arguments can be made drawing certain boundaries, setting certain inputs to zero, letting the contribution of certain subsystems be thought of as constant over relevent time spans, and the like.

With a secularized (ultimately dialectical) epistemology at our command, with a variety of models to exploit in a flexible way in a search for robust results, and with many different closure options to explore, there is no reason why game theory cannot be a useful tool (Levins, 1966; Lewontin, 1976). It ceases to be a useful tool when it is used one-dimensionally in terms of crypto-teleological strategies like those attributed to the rational economic species.

An interesting comparison that illustrates the last point nicely is the following. Bookkeeping entries, personal accounts, or patterns of behavior which apparently indicate that a human agent is behaving as a rational economic man are, in fact, compatible with a wide range of explanations, including the presence of structural constraints of all sorts. No immediate inference is available from the bookkeeping of atomistic rational economic behavior to explanations for the behavior. An evolutionary parallel to this point has been made amply and elegantly by Elliot Sober and others (Sober and Lewontin, 1982; Gould and Lewontin, 1979). Namely, the bookkeeping of population statistics and the apparent adaptational utility of particular traits are not sufficient to allow us to conclude that those particular traits were selected for, in the standard neo-Darwinian sense, though they arose through normal evolutionary processes. Many other potential explanations are available, consistent with the body of bookkeeping data.

The temptation in each case to make the immediate inference from bookkeeping data to simplistic explanation in terms of the ideal types rational economic man or Darwinian selection betrays an a priori commitment to those ideal types that is not defensible on any scientific grounds.

Explicitly stated in terms of game theory, this means that a matrix seeming to yield stable optimal strategies in a single game may, in fact, be the consequence of suboptimal satisficing in a complex array of interpenetrating games played by players at several systems levels, sometimes deterministically, sometimes stochastically, all at the same time or during overlapping spans over evolutionary time. The closure condition favoring the explanation in terms of the simple one-level game is usually imposed by brute force, and explicitly bars us from investigating the more complex possibilities.

Closure and Method: The Laboratory and the World

In a very dialectical way, the most successful scientific strategies show up the problem of closure best. Experimental design is largely a techique for imposing closure on systems so that clear-cut explanatory connections can be established. When experimental design is at its best it creates the widest gap between a phenomenon contrived in the laboratory and the phenomena found in a natural setting. In the laboratory, subsystems are artificially sealed off, and their isolated behavior is studied. It takes a very complex set of inferences to move from what is learned in the laboratory to an understanding of what is going on in nature. The complexity of the move is seldom appreciated. By ignoring the complexity, and by thinking of the natural (and social) world as an enormous laboratory, the results of the narrow neo-Darwinian program can be made to look more impressive than they really are.

A certain sensitivity to the presence of the gap between laboratory and nature is shown by John Thompson in a work already cited. Two passages in particular are directly related to my emphasis on explanatory closure. In the first, Thompson refers to laboratory experiments investigating inter- and intraspecific competition among snails. He says:

In the laboratory experiments, interspecific competition and its effect on growth of snails was as intense as intraspecific competition, thereby indicating that interspecific competition could be an important selective force. Also, individuals of the same size ingested the same size particles in all Hydrobia species, and individual growth of snails was correlated with the availability of diatoms of particular sizes. Together these results indicate that species with the same size frequency distribution have nearly complete overlap in sizes of the food they select and they compete for the resources. What is missing from the analysis, of course, is a demonstration that food is limiting in natural populations. Also the extent to which the populations mix from year to year is unknown (Thompson, p. 42).

The two concluding sentences are the key from my point of view. The first of them explicitly states the need for straightforward Malthusian closure if we are to extend the laboratory results to nature. In fact, of course, every selectionist explanation based on the competition for scarce resources depends on the demonstration that resources are indeed limitingly scarce. Oftentimes this sort of closure is attempted by assuming that populations will naturally expand to the point of Malthusian closure. But this is a bizarre assumption. For in turn it assumes that there are no constraints in any other relevant dimension or at any other relevant level that keep the population below Malthusian limits in the dimension being focused on. Now I have no doubt that laboratory conditions can be controlled in such a way that laboratory populations are under Malthusian closure with respect to given resources. There is, moreover, little doubt that some populations in the wild are at Malthusian limits with respect to some given resource. But I seriously doubt that Malthusian closure ever exists for any population in all dimensions at any time. To assume so would be to assume that the earth is a thermodynamic plenum, a zero-sum game with respect to all accounting systems. But the earth is a thermodynamically open system as a whole, as are all ecological subsystems.

It seems probable to me that when Malthusian closure does occur it occurs not as a result of population expansion, but more often as a result of environmental contraction as a consequence of climatological and geophysical events.

We can accept selectionist explanations based on competition for scarce resources only when Malthusian closure can reasonably be demonstrated. It seldom is. What usually happens is the reverse. We are told that when plausible selectionist explanations can be provided, then we can reasonably assume that Malthusian closure was present. This of course is totally question-begging unless we have an a priori commitment to the exclusiveness and ubiquity of such explanations. The so-called "Just-So Stories" that have become famous in the literature of evolutionary biology, and especially sociobiology (Gould and Lewontin, 1979) are usually explanation guesses without the necessary closure conditions having been established. Somewhere along the line it was apparently decided that Mother Nature is a good frugal bourgeois Hausfrau.

One of the consequences of good—especially good—experimental design is to seal out interaction effects, as Thompson also points out. Failure to do so results in failure to achieve a cleanly interpretable experiment. This is far from a criticism of experimental methods. It is rather a clear statement of their simultaneous advantages and limitations. Let us listen again to Thompson:

Most of what is interesting about biological communities cannot be pinned, stuffed, pressed onto herbarium sheets, or preserved in alcohol. Knowing the species structure of an assemblage of organisms tells us in and of itself little more than a telephone book tells us about a city. Nor can what is interesting about biological communities be dissected, weighed, separated on starch gels, or centrifuged into supernatant and precipitate fractions. Knowing the internal workings of organisms in isolation from other organisms with which they interact tells us the "how" of life without the "why". What makes biological communities more than lists of taxa complete with details of how they tick are the interactions among species (Thompson, p. 124).

Again this is a warning about hasty inferences from lab to nature. We are confronted with the question of how experimental findings are to be integrated into explanations of natural phenomena. One key to answering the question is to push Thompson's distinction between the how of life and the why. Laboratory experiments tell us how things do happen in the lab, and how they could happen in nature.[6] Any attempt to demodalize experimental findings for application to natural situations requires us to make a dialectical return to the conditions imposed on the experiment itself in order to recall the imposed closure which made the experiment possible in the first place. An assessment of these closure conditions is essential to any claim that nature is like the laboratory in exactly the way it must be if the inference from lab to nature is to be acceptable. Thompson's distinction between how and why is very like the distinction I've been making between bookkeeping and explanation.

In terms I have been using throughout this chapter, Thompson's book as a whole constitutes an argument against the inference from any one experiment to explanation within evolutionary ecology. Nor can experiments simply be added together. Every experiment will impose closure, limit interactions, etc., in a particular way. The assumptions allowing us to say that any graphs derived from two experiments can be combined in a simple way are likely to be very restrictive ones and, of course, have to be justified with respect to their fit with natural circumstances as well as with one another.[7]

The argument of this chapter is that complexity has to be confronted honestly, and that orthodox neo-Darwinian explanation patterns often fail to do so. This does not minimize the debt that we owe to the orthodox Darwinians, nor deny the fact that orthodox Darwinian explanations are bound to have a place in any more comprehensive theory, but it does limit the importance of orthdox selectionist explanation. The reply by the orthodox neo-Darwinian to this line of criticism could well be to admit the complexity of evolution, but argue that this complexity can be dealt with by taking ordinary natural selection as the standard explanatory model, and then treating all additional complexities as modifiers of the underlying standard Darwinian events. Unfortunately, this strategy runs into all the problems of ideal types and ceteris paribus clauses we have canvassed. In addition, an argument to this effect is usually forced to have recourse to simplicity as one of the values (Kuhn, 1962, 1970) upon which theory choice is to be based. In other words, at some stage, anyone who insists on complexity will be confronted with Occam's razor.

As it is normally used now, the Occam's razor argument comes down to a scruple. We are to choose between competing theories (ceteris paribus) by favoring the simpler one—assuming that the simpler one has an adequate track record as an explainer and, since Lakatos (1970), as a generator of new research. Put this way, the scruple sounds like a sensible heuristic. In fact, however, as many have pointed out, it is problematic in devastating ways.

First we have to remember that simplicity is inextricably theory-bound, and hence that Occam's razor always depends on a criterion internal to a particular theory. The simplest explanations for the phenomena we like to explain by evolutionary theory are the explanations preferred by some freshmen: God made it (us) that way. When the teacher asks how God did it, he is often told that's God's business and we ought to confine ourselves to what is proper for us to know. There is a long dialectical sequel to this *aporia*, and a short one. The long one involves spelling out the metaphysical and/or existential commitments underlying the two opposing views. No matter how the dialogue unfolds, the invocation of Occam's razor must be absolutely question-begging.

The short sequel consists of the teacher leveling the charge of unscientificness against the student. But this is a dangerous move, for the student must be antecedently impressed by the wonders of science if the charge is to have any persuasive power. Only when the power of the charge is already acknowledged (e.g., when the student's dependence on the fruits of scientifically grounded technology is established) can Occam's razor be invoked. But then the key move is not the razor stroke. The key is the entrapment of the student in an internal commitment to science. This is not just an anecdote about freshmen, of course. It is a microcosm of the entire creationism/evolutionism debate.

The competing theories sketched in this chapter as alternatives are far less globally antithetical than the faith-versus-reason alternatives argued by teachers and freshmen. Yet appeals to Occam's razor are equally question-begging. One-dimensional selection models seem simple to selectionists, but dressed up in qualifying caveats until only their own mother could recognize them, their simplicity is far from obvious to the unfaithful. Every model is a candidate for simplicity in its own terms. The real issue is which models, handled with which epistemological strategies, are the better ones in terms of their adequacy as explainers and research generators. None of the sides in a dispute between theories can insist on the a priori authoritative stance that would be required in order for Occam's razor to have any persuasive force.[8]

Notes

1. A challenging, perhaps infuriating, account of *McLean* is contained in Geisler (1982). Despite its partisanship it contains the important documents pertaining to the case; and because of its partisanship ought to be required reading for anyone wishing to defend evolutionary theory in the public forum.

2. Thus Conant (1947/1951) says that were he to be teaching a course whose aim was for the layman to understand science, "I should wish to show the difficulties which attend each new push forward in the advance of science, and the importance of new techniques: how they arise, are improved, and often revolutionize a field of inquiry. I should hope to illustrate the intricate interplay between experiment, or observation, and the development of new concepts and new generalizations; in short, how one conceptual scheme for a time is adequate and then is modified or displaced by another. I should want also to illustrate the interconnection between science and society about which so much has been said in recent years by our Marxist friends" (Mentor edition, pp. 31–32).

3. My conviction that a focus on closure conditions is essential has two sources: (a) the work of Roy Bhaskar (1975 and 1979); (b) a line of thought developed by Prigogine and Stengers (1984); Pattee (1970, 1972, 1973, 1978); Weiss (1973); and others.

4. For example, "Often the species that have strong effects on the population of other species will be the same as the species critical in the evolutionary unit of interaction, but this will not always be the case. For example, some mutualisms may have no effect on the population levels of interacting species. A mutualism is favored by selection because it allows those individuals possessing traits that foster the interaction to increase their genetic contribution to future generations relative to other individuals in the population. The mutualism may have little or no effect on the overall population levels of the species. Experiments designed to determine the unit of interaction within which selection acts significantly on all the species cannot use changes in population levels resulting from a manipulation of one of the species as the sole criterion of the limits of the important species" (Thompson, 1982, p. 126).

5. Issues such as these are discussed with great care in virtually all the papers collected in Futuyma and Slatkin (1983). Apropos the present point the editors say "The study of coevolution forces a different view of genetic evolution than is usually adopted. In population genetics and evolutionary theory, each species is usually considered in isolation, with the environment and associated species relegated to the background, which is assumed to remain unchanged. Co-evolutionary theory . . . assumes that genetic changes may occur in all interacting species, allowing genetic changes to be driven both by immediate interactions and by the feedback through the rest of the community. The distinctive feature of coevolution is that the selective factor (a predator) that stimulates evolution in one species (a prey) is itself responsive to that evolution, and the response should be predictable. In some cases a coevolutionary equilibrium may be

established. In other cases there may be no coevolutionary equilibrium, and evolution may continue over longer time scales than are typical for the attainment of gene frequency equilibria as usually treated in population genetic models" (p. 6).

6. Compare Futuyma and Slatkin (1983, p. 9): "Pimentel and his coworkers (Pimentel and Stone, 1968) have found evidence of genetic changes in both houseflies (*Musca*) and the parasitoid wasp *Nasonia vitripennis* when cultured together, and Hassell and Huffaker (1969) reported increased resistance in the host and increased effectiveness of the parasitoid in a moth-wasp laboratory system. Such studies show, of course, that pair-wise coevolution is possible, not that it commonly occurs in nature. In the absence of an actual history of the dynamics of genetic change, the demonstration that each of two interacting species is genetically variable for the characteristics that affect their interaction can at least show the potential for coevolution."

7. Thus the analogy between breeding practices and natural selection, so frequent in *The Origin of Species*, may be more misleading than Darwin thought. There may be significant differences between the evolutionary dynamics of Darwin's finches and those of Darwin's whippets.

8. Work was begun on this chapter while I was attending the 1982 Summer Institute of Philosophy held at Cornell University, sponsored by the Council for Philosophic Studies, and supported by NEH. I thank Marjorie Grene and Dick Burian for allowing me to attend. David Depew and John Jungck provided essential help during the writing. As usual, the collaboration of Grace Stuart doubled the area under my production possibility curve. Work was completed on a Research and Study Leave grudgingly granted to me by Temple University.

References

Allen, T. H. F., and T. B. Starr, 1982. *Hierarchy: Perspectives for Ecological Complexity*. Chicago: University of Chicago Press.

Bhaskar, R., 1975. *A Realist Theory of Science*. Atlanta Highlands, New Jersey: Humanities Press.

Bhaskar, R., 1979. *The Possibility of Naturalism*. Atlanta Highlands, New Jersey: Humanities Press.

Beatty, J. H., 1980. What's wrong with the received view of evolutionary theory? In *PSA 1980*, vol. 2, P. Asquith and R. Giere, eds. East Lansing, Michigan: Philosophy of Science Association, pp. 397–426.

Bourdieu, P., 1977. *Outline of a Theory of Practice*. Cambridge: Cambridge University Press.

Brandon, R., 1978. Adaptation and evolutionary theory. *Stud. Hist. Phil. Sci.* 9:181–206.

Burian, R., 1983. Adaptation. In *Dimensions of Darwinism: Themes and Counter Themes in Twentieth Century Evolutionary Theory*, M. Grene, ed. Cambridge: Cambridge University Press.

Conant, J. B., 1947/1951. *On Understanding Science*. New Haven: Yale University Press.

Dawkins, R., 1976. *The Selfish Gene*. New York: Oxford University Press.

Feigl, H., 1958/1967. *The "Mental" and the "Physical"*. Minneapolis: University of Minnesota Press.

Feyerabend, P. K., 1975. *Against Method*. London: Verso.

Foucault, M., 1970. *The Order of Things: An Archeology of the Human Science*. New York: Vintage/Random House.

Futuyma, D. J., and M. Slatkin, 1983. *Coevolution*. Sunderland, Massachusetts: Sinauer Assoc.

Gatlin, L. L., 1972. *Information Theory and The Living System*. New York: Columbia University Press.

Geisler, N. L., 1982. *The Creator in the Courtroom*. Milford, Michigan: Mott Media Inc.

Gillespie, N. O., 1979. *Charles Darwin and the Problem of Creation*. Chicago: University of Chicago Press.

Gould, S. J., and R. C. Lewontin, 1979. The spandrels of San Marco and the Panglossian paradigm. *Proc. Roy. Soc. London B.* 205:581–598.

Hamilton, W. D., 1964. The genetical theory of social behavior, I and II. *J. Theoret. Biol.* 7:1–52.

Kuhn, T. S., 1962/1970. *The Structure of Scientific Revolutions*. Chicago: University of Chicago Press.

Lakatos, I., 1970. Falsification and the methodology of scientific research programmes. In *Criticism and The Growth of Knowledge*, I. Lakatos and A. Musgrave, eds. Cambridge: Cambridge University Press.

Lakatos, I., and A. Musgrave, 1970. *Criticism and the Growth of Knowledge*. Cambridge: Cambridge University Press.

Levins, R., 1966. The strategy of model building in population biology. *Am. Scientist.* 54:421–431.

Levins, R., 1968. *Evolution in Changing Environments*. Princeton: Princeton University Press.

Lewontin, R. C., 1976. Evolution and the theory of games. In *Topics in the Philosophy of Biology*, M. Grene and E. Mendelsohn, eds. Dordrecht: Reidel, pp. 286–311.

Margolis, J., 1983. *Philosophy of Psychology*. Englewood Cliffs, New Jersey: Prentice-Hall.

Mayr, E., 1977. Darwin and natural selection. *Am. Scientist* 65:321–327.

Mills, S. K., and J. H. Beatty, 1979. The propensity interpretation of fitness. *Phil. Sci.* 46:263–286.

Pattee, H. H., 1970. The problem of biological hierarchy. In *Towards a Theoretical Biology*, C. H. Waddington, ed. Chicago: Aldine, pp. 117–136.

Pattee, H. H., 1972. The evolution of self-simplifying systems. In *The Relevance of General Systems Theory*, E. Laszlo, ed. New York: Braziller, pp. 31–41.

Pattee, H. H., 1973. *Hierarchy Theory*. New York: Braziller.

Pattee, H. H., 1978. The complementarity principle in biological and social structures. *J. Soc. Biol. Structures* 1:191–200.

Prigogine, I., and I. Stengers, 1984. *Order Out of Chaos*. New York: Bantam.

Ruse, M., 1971. Natural selection in the *Origin of Species*. *Stud. Hist. Phil. Sci.* 1:311–351.

Ruse, M., 1975. Charles Darwin's theory of evolution: an analysis. *J. Hist. Biol.* 8:219–241.

Simon, H. A., 1981. *The Sciences of the Artificial*. Cambridge, Massachusetts: The MIT Press.

Slobodkin, L. B., 1961. *Growth and Regulation of Animal Populations*. New York: Dover.

Slobodkin, L. B., 1968. Towards a predictive theory of evolution. In *Population Biology and Evolution*, R. Lewontin, ed. Syracuse: Syracuse University Press.

Sober, E., and R. C. Lewontin, 1982. Artifact, cause, and genic selection. *Phil. Sci.* 49:157–180.

Thompson, J. M., 1982. *Interaction and Coevolution*. New York: Wiley.

Tuomi, J., and E. Haukoija, 1979. An analysis of natural selection in models of life-history theory. *Savonia* 3:9–16.

Tuomi, J., Jukka Salo, Erkki Haukioja, Pekka Niemelä, Tuomo Hakala, and Rauno Mannila, 1983. The existential game of individual self-maintaining units: selection and defense tactics of trees. *Oikos*. 40:369–376.

Turner, J. R. G., 1983. "The hypothesis that explains mimetic resemblence explains evolution": the gradualist-saltationist schism. In *Dimensions of Darwinism*, M. Grene, ed. Cambridge: Cambridge University Press, pp. 129–169.

Weber, M., 1947. *The Theory of Social and Economic Organization*. Oxford: Oxford University Press.

Weiss, P. A., 1973. *The Science of Life*. Mt. Kisco, New York: Futura Publishing.

Wilson, E. O., 1971. *The Insect Societies*. Cambridge, Massachusetts: Harvard University Press.

An Organizational Interpretation of Evolution

John H. Campbell

The structure-function principle defines biological explanation and sets biology apart from the physical sciences. Organisms carry out biological processes as functions of their adaptively evolved structure. From early gross anatomy to modern molecular biology, structure has been sought to rationalize function. Some fields of biology have yet to achieve this level of explanation because the underlying structures remain too inaccessible for rationalizing the concept of function to be useful. Evolution is a case in point. When Darwinism and even modern neo-Darwinism were formulated, the structures of organisms and genes were so vaguely known that evolutionists deliberately denied their roles to avoid the specter of vitalism. Instead, evolutionists reverted to the mechanical paradigm of physics, in which inert objects move only in passive response to exogenous forces pushing upon them from the outside. Evolution became change that the external environment forces upon the hapless species instead of a function that organisms are structured to carry out. In this Darwinian perspective, species do not evolve in an active sense; they only get evolved by an external natural selector. They do not have causal roles in their evolution.

During the past decade, molecular biology has advanced genetics to a structure-function science. This progress is forcing us to recast evolution into a truly biological science as well. The profound new discoveries for evolution no longer concern the demands of the environment on species, but rather the molecular structure of organisms themselves. These discoveries raise profound questions as to how biological structures function to direct, cause, and carry out evolution.

Complex Structure of the Genes

Studies of gene structures have brought to light three organizational features that have not been incorporated into evolutionary theory. First,

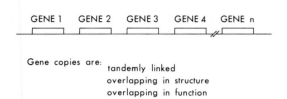

Figure 1
Basic organization of a multigene family.

genes have significant internal structure. This substructure is elaborate, unique to each gene, and the basis for individual gene function. For example, most types of proteins in mammals are not represented by a single coding sequence of DNA, as we had once supposed (Flanagen and Rabbits, 1982). Instead, they are encoded by a series of related gene copies organized as multigene families (Hood, Campbell, and Elgin, 1975).

Figure 1 illustrates a basic multigene family, although they can be specialized in surprisingly diverse ways for particular functions (Campbell, 1983). The example contains an array of gene copies that are interrelated by being (1) linked together in tandem, (2) homologous in base sequence, and (3) overlapping in function. In some families the gene copies are identical to one another. In others they differ to some degree in structure or function as appropriate for the phenotype that they specify. Multigene family organization enormously increases the capacity of DNA to store, process, and express genetic information. It also allows—and demands—evolution to proceed quite differently than for simple, single-copy genes. For example, when two or ten or a thousand similar gene copies are tightly linked together, selection becomes unable to operate on mutations in a single copy. The individual gene is lost in the crowd, and natural selection is able to survey only the adequacy of the family as a whole.

The second realization about genes is revolutionizing both our ideas and our practice of genetics. Genes are chemical substrates on which enzymes operate. The early view of genes as static units of information faithfully handed down passively from generation to generation, from mother to daughter cell during development, is inaccurate. Genes are not sacred messages but profane molecules. As such, their structures can be and are deliberately altered by the organism (Campbell, 1982). Cells are replete with diverse enzymes capable of effecting every imaginable sort of alteration in DNA structure. Here is a list of some enzymes that process genes:

Back transcriptase	Acquisitionase
Correcting enzymes	Excisase
DNA gyrase	Insertase
DNA ligase	Integrase
DNA methylase	Invertase
DNA polymerase	Mutase
DNA topoisomerase	Recombinase
Glycosylase	Replicase
Helicase	Spligase
Primase	Translocase

These gene-processing enzymes make remarkable alterations in the structure of gene molecules, and are essential to the process by which elaborate genes, such as multigene families, evolve. For example, enzymes expand and contract certain multigene families, shoot point mutations into selected reiterated gene copies, correct the structure of one gene copy against another, splice together segments of various gene copies into new recombinant genes, scatter individual gene copies from families elsewhere in the genome, transfer silent gene copies into special sites at which they can be expressed, excise and discard entire multigene families from a chromosome, and so forth (Campbell, 1983). These enzymatic activities are indispensable for the expression of complex multigene families and to their evolution as well. Needless to say, enzymatic alterations of DNA molecules are carried out in precise and controlled ways so as to serve useful roles instead of creating chaos. To this end, multigene families have evolved elaborate control systems called governors (Campbell, 1983) which regulate their structural alteration by enzymes.

Gene-processing enzymes make the genome far more fluid and dynamic than had been imagined earlier (Hunkapiller et al., 1982). They probably are the source of most of the genetic variability important for evolution. This is significant because enzymes are notable for the specificity of the chemical transformations that they catalyze, while the hallmark of classical mutation is its presumed randomness. Profane

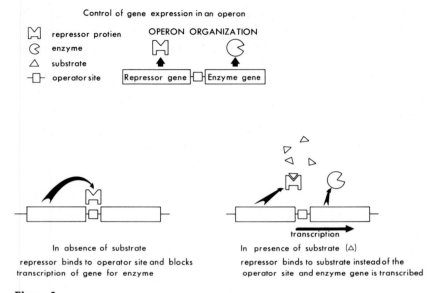

Figure 2
Basic operation of a bacterial operon. The gene for an enzyme is expressed only when its substrate is present through control by a repressor protein which acts as a sensor for substrate molecules. The repressor blocks transcription in the absence of substrate. Substrate inactivates the repressor and allows gene expression.

gene molecules thus offer patterned variation for natural selection. Even those mutational changes not actually catalyzed by enzymes are picked over and winnowed by multiple systems of sophisticated error-correcting enzyme pathways to leave a small pool of highly edited mutant alleles for selection.

The third profound realization of modern genetics is that complex genes include sensing devices to bring relevant information from the environment to their DNA molecules. Figure 2 illustrates the basic structure of an operon (Jacob and Monod, 1961), a common type of sensing device among bacterial genes. The organization shown allows a gene for an enzyme to be expressed only when substrate for the enzyme is present. This conditionality is provided by a second protein called a repressor which is encoded next to the gene for the enzyme. The repressor can attach specifically to the DNA at the start of the gene and block its transcription. Alternatively, the repressor can bind specifically to substrate molecules for the enzyme if they are present. Since the two binding activities are mutually exclusive, the presence of substrate dislodges the repressor from the DNA and opens the enzyme gene for expression. Geneticists say that the presence of substrate "induces" the expression of the gene (Monod, 1957). Repressor systems are useful (i.e., increase fitness) because they allow bacteria to make enzymes only when appropriate. For example, inducibility can relieve the cell from the burden of synthesizing massive quantities of an enzyme when there is no substrate for it to act upon.

Considering their molecular simplicity, repressors are sensing devices of remarkable versatility. In theory a repressor could evolve to inform a gene about any ligand or condition able to affect the quaternary structure of a protein. While sensors still are known best in prokaryotes, they are also integral components of eukaryotic control systems, such as multigene family governors.

Thus genetics has become a science of structure-function relationships. Of course, many genes have only simple structure and behavior, but these tend also to have correspondingly trivial functions. Complex functions, on the other hand, are specified by highly organized genetic determinants. These are the genes important to evolution. Just mapping and counting their mutant alleles is not enough, either for the current geneticist or the evolutionist. The commitment of modern molecular biology is to determine the exact structure of important genes and to explain mechanistically how particular activity flows from particular structure. This new era of genetics is disclosing a remarkable new type of biological function. Some genetic structures do not adapt the organism to its environment. Instead, they have evolved to promote and direct the process of evolution. They function to enhance the capacity of the species to evolve.

Genes with Evolutionary Functions

Since evolutionary function is not a familiar concept, it is useful to describe a simple concrete example. The genes that are most important for bacteria to adapt to their immediate environment typically are organized into special structures called transposons (Kleckner, 1981). Figure 3 is a diagram of a typical example, called Tn3, coding for resistance to penicillin in Enterobactereacae (Chou et al., 1979). This transposon is a segment of DNA encompassing the gene for the enzyme beta-lactamase, which hydrolyzes penicillin. Its two ends are marked with a duplicated recognition sequence. These are target sites for a transposase enzyme, which can move the unit to other regions of the chromosome.

STRUCTURE OF TRANSPOSON   Tn3

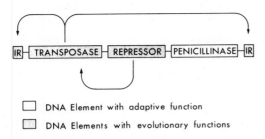

☐   DNA Element with adaptive function

▨   DNA Elements with evolutionary functions

Figure 3
Structure of a penicillin-resistant transposon, Tn3. Tn3 is a segment of DNA encompassing several genetic elements. The penicillinase gene has an adaptive function of specifying the drug-resistant phenotype. The transposase, repressor for the transposase gene and IR ("inverted repeat") target sites for transposase, have the evolutionary function of transposing the Tn unit about the genome.

The transposon also codes for two other proteins. One is the transposase. The second has several functions, including that of a repressor to regulate expression of the transposase gene (Grindley, 1983). Thus the transposon literally transposes itself through the machinery that it encodes and controls.

Transposons are now recognized to be foremost agents for bacterial evolution (Anderson, 1968; Davey and Reanney, 1980). In particular they allow genetic information to hitchhike on special larger structures called transmissible plasmids from cell to cell, and even from species to species in the ecosystem. Bacteria adapt to alterations in environmental conditions, both beneficial and deleterious, primarily by capturing transposons.

Transposons also abound in eukaryotes. Here they may have various evolutionary roles (Temin and Engels, 1983) especially related to genomic rearrangements and to speciation (Spradling and Rubin, 1981). Some also may be involved in cellular differentiation (Davidson and Britten, 1979) or various other specialized adaptive functions (McClintock, 1965).

One should note that the structures of Tn3 surrounding the beta-lactamase gene do not increase the fitness of the host organism. A bacterium with only the beta-lactamase coding sequence is just as resistant to penicillin as one with the complete transposon. The function of transposition mechanism is to enhance the ability of the species to evolve. This evolutionary function is distinct from the adaptive function of protecting the cell from penicillin, and the two are carried out by distinct structural elements of the transposon.

How Do Species Evolve Structures for Evolutionary Function?

Traditional Darwinism provides a simple and powerful mechanism to evolve structures with adaptive function. The individual organisms best able to carry out adaptive functions automatically have selective advantage in propagating their particular genes. Selection for evolutionary functions is not that direct. Structures that help the species to evolve do not increase the competitive fitness or fecundity of the individual. Nevertheless, such structures undeniably are able to evolve, as shown by the elaborate structure of Tn3. A clue as to how this evolution occurs lies in a dynamic aspect of the evolutionary process called the Red Queen paradox by Van Valen (1973). The reference is to the queen in *Alice Through the Looking Glass*, who told Alice as they raced at top speed not to expect to "get anywhere" by running. Inhabitants of Looking Glass Land were obliged to run as fast as they could just to stay where they were. To get somewhere they would have to run "ever so much faster."

Species in an ecosystem are caught in this dilemma. Their access to resources depends upon their ability to compete with the other species in this environment. A relatively well-adapted species will be able to command a greater share of the ecosystem's resources (food, shelter, and so forth) but only through competition with neighboring species. Van Valen realized that the challenge in the ecosystem is more than just to be fit. More ominously, competitors are also evolving. Being fit is sufficient in the short term, but in the long run a species can maintain its status in the ecosystem only by continuously matching the adaptive inroads made by its predators, prey, and competitors. This ceaseless adaptation does not improve a species' ecological position. It is a requirement just for maintaining the status quo. To get somewhere ecologically—to expand in population size, to radiate, and to predominate in new niches as they develop—a species must evolve "ever so much faster."

In the long run competition is not just for fitness with the environment but is for the ability to out-evolve other species. Those forms most facile at evolving will predominate. Ultimately, evolutionary success for each competitor comes from acquiring tricks, skills, and strategies to evolve faster and more effectively than the competition. These are

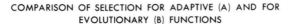

COMPARISON OF SELECTION FOR ADAPTIVE (A) AND FOR
EVOLUTIONARY (B) FUNCTIONS

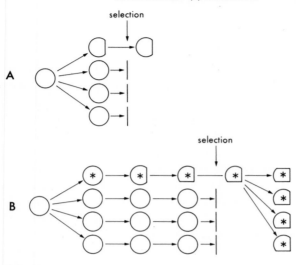

Figure 4

Comparison of selection for adaptive (A) and evolutionary (B) functions. A: The col-
umns of figures represent competing individuals or populations. An alteration in
adaptive phenotype, indicated by diagrammed shape, can be directly selected for or
against. B: * represents a genetic alteration that enhances the capacity of the population to evolve but does not in itself affect fitness. With time this capacity promotes
adaptively significant phenotypic changes that can be selected for. Both types of evolutionary change occur through selection of genetic changes on the basis of the consequences of those changes.

capacities that phenotypic and genotypic organization can bestow. Species that develop phenotypes that enhance their potential for subsequent
evolution eventually will be favored by the results of that potential.
Structures that interfere with meeting subsequent evolutionary challenges will eventually be winnowed. Thus species can develop an ability
to evolve through the same fundamental principle of "selection by
consequences" (Skinner, 1981) that underlies adaptation.

Figure 4 compares the operation of selection in these two evolutionary
processes. The main difference between the two is that it takes longer
to select genes for evolutionary functions. This difference does not
present any philosophical problems (Layzer, 1980) although it does
raise the practical question whether structures can acquire useful evolutionary roles fast enough to be significant factors in evolution.

Fortunately, a number of special mechanisms can speed up the evolution of elaborate structures for evolutionary function. The most important of these mechanisms is *evolutionary recruitment*. Here a phenotypic trait develops through direct natural selection because it helps to adapt the species to its environment. This trait is then modified so that it also performs an evolutionary function. Just one or two mutations may recruit a very complex preformed adaptation into an evolutionary role. The mutations are favored because the lineage of organisms that acquires them becomes better able to evolve. In this way the sophisticated structures that evolve through the powerful process of adaptation serve as preadaptations for evolutionary functions.

To illustrate evolutionary recruitment, let us return to bacterial transposons, but this time to an erythromycin-resistance determinant of *Streptococcus*, Tn917. This transposon has properties similar to those described earlier for Tn3 but with a couple of extra elaborations (Tomich, An, and Clewell, 1978). It has a repression system so that the resistance gene is expressed only in the presence of erythromycin. Presumably the repressor evolved adaptively as discussed above for other operons. In addition, Tn917 transposes only when erythromycin is present. The repressor serves a second, evolutionary function of rendering the transposon mobile only when it is relevant to the phenotype. It is implausible that the modest evolutionary benefit provided by controlling transposability was adequate to create an erythromycin-sensitive repressor from scratch. However, once a repression system evolved adaptively, its influence was easily extended to the transposase gene. Figure 5 diagrams this suggested pathway of evolutionary recruitment.

Evolutionary Drivers

Biological structures that serve a particular adaptive function are called adaptations. It is useful to coin a different name for structures that serve evolutionary functions. I suggest the term *evolutionary drivers*. Evolutionary drivers are units of biological structure that carry out evolutionary functions instead of adapting the organism to its environment. We can now parphrase the preceding section by saying that evolutionary recruitment is the process of developing an adaptation into an evolutionary driver.

Evolutionary drivers are distinct from adaptations (Layzer, 1980), although it is possible for a structure to have both functional roles simultaneously. In particular, when an adaptation is recruited to become

Evolutionof a repressor withboth an Evolutionary and adaptive function
in a transposon

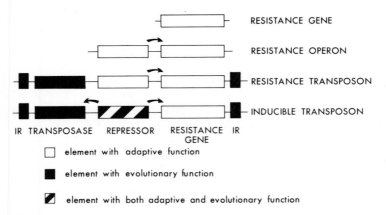

Figure 5
Proposed pathway for the evolution of a conditionally transposable antibiotic resistant
determinant, such as Tn917, by recruitment. Four evolutionary stages are represented.
An initial simple antibiotic-resistance bacterial gene is converted into an operon
through the adaptive evolution of a repressor that senses the presence of the drug.
The operon is then translocated, by mutation, into a transposable segment of DNA
that codes for its transposase. Finally, a regulatory mutation extends the influence of
the repressor to the transposase gene, giving it both an evolutionary and an adaptive
function.

an evolutionary driver, it may retain its former adaptive function as
well. As mentioned above, an erythromycin-sensitive repressor controls
both the expression of the transposase gene and the resistance gene
of Tn917. Such duality of function is another mechanism that can
facilitate the evolution of an evolutionary driver. As the structure is
modified to meet changing selective conditions, its specificity as an
evolutionary director is also updated.

I have chosen the term *driver* because it implies the internal source
of the influences on the species' evolution. It also covers two types of
ways that evolutionary drivers operate in evolution. Some drivers provide direct forces for evolutionary change. Their effects have been
called *drive* for a quarter of a century: meiotic drive, chromosomal drive,
gametic drive, mutational drive (Hiraizumi, Sandler, and Crow, 1960;
Lyttle, 1977) and more recently, molecular drive (Dover, 1982).

The second, and perhaps more important role of biological structure
is to direct or channel the pathway of evolution without actually power-

ing the change. Such structures drive evolution in the same sense that a bus driver drives his vehicle. Since Darwinian selection amalgamates these two roles, they are frequently taken to be inseparable by evolutionists. Therefore, it is worth considering in detail how biological structure directs evolution without fueling it. To remove any ambiguity in the discussion, I shall use the term *evolutionary director* for those biological structures which only steer evolution.

An important realization is that the environment presents the species with a challenge, but the species itself determines which way it responds to that challenge. Different species will respond to a common environment in different ways. This role of the species involves an important aspect of evolution that is too poorly appreciated by modern neo-Darwinists—the existence of alternative pathways for adaptive evolution. In response to any environmental stress or change, species may have the potential to adapt in many different ways. The structure of the species determines which pathway it will take.

The shapes of horns on antelopes pleasingly illustrate a diversity of equivalent pathways that evolution has taken to meet one environmental challenge. Each of the twenty of so species of African antelopes has a recognizably different horn (Simpson, 1950). Simpson carefully considered the question of why the horns of various species evolved differently, and concluded that the differences were not adaptively significant. Instead the various forms seem simply to be alternatives. Each species was called upon to evolve horns; a horn must have some shape, and so each species evolved a horn that was adequate. Simpson suggested that the particular pathway of evolution taken by each individual species was determined by the types of genetic variability available for selection in that species' gene pool, rather than by the environment.

Antelopes nicely illustrate the fact that evolution can go in different pathways. But higher organisms are far too complex for analyzing how a species opts among possible alternatives. The genetics of antelopes are unknown, as are the exact multiple functions of their horns. Also, antelopes are not suitable for long-term evolutionary experiments. On the other hand, certain traits of bacteria are particularly well understood, and it is possible to observe experimentally how the structure of the species affects their evolution. For example, one can easily maintain 500 replicate populations of bacteria under identical conditions and record their evolution over thousands or even tens of thousands of generations. I will not describe such experiments that I and others have done on bacterial evolution but merely summarize some conclusions relevant to evolutionary directors:

• A species of bacteria can adapt to a change in the environment, such as the introduction of a novel energy source, in multiple, distinct ways (Langridge, 1969).

• Many of these adaptive changes are alternatives to one another in that (a) any of them will adapt the species and hence is selectively permissible; (b) acquiring one change reduces or eliminates the selective forces for acquiring a second change; (c) when replicate populations of the species are propagated in the same environment, some will adapt by one type of change and the others by the alternative changes.

• The relative likelihood that a population will take a particular evolutionary pathway instead of an alternative can be measured with replicate populations and expressed as an option coefficient.

• The values of option coefficients depend upon the genotype of the strain. By starting with characterized mutant strains, one can determine which genetic determinants influence the value of an option coefficient. These genes code for evolutionary directors.

• Some evolutionary directors bias a species to adapt in one general sort of way rather than another for a range of different environmental factors; for example, to evolve by gene duplication instead of by nucleotide base substitutions or to alter one aspect of phenotype in preference to another. These directors should bias the species over long periods of evolutionary time to evolve according to a pattern of genotypic change.

• If a pattern imparted to evolution by an evolutionary director has a cumulative directional component, the evolutionary director will cause long-term evolution to progress in a definable direction (Sueoka, 1965).

• Evolutionary directors are genetically determined characteristics of a species and therefore can change and evolve. If a director mutates, a species can change from evolving in one pattern or direction to another. (For an example in an experimental system see Cox and Yanofsky, 1967.)

• Selecting for a change in phenotype generates an attendant selection for evolutionary directors that favors that particular pathway of change.

This last point is important enough to warrant an example, as in figure 6. When one selects for mutations in bacteria, one concomitantly selects for cells with increased rates of that type of mutation. This is

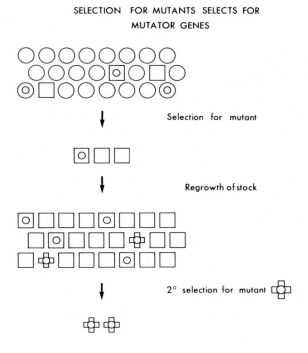

SELECTION FOR MUTANTS SELECTS FOR
MUTATOR GENES

Selection for mutant

Regrowth of stock

2° selection for mutant

O   Cell with genetically determined increased
mutation rate
Shape of cells indicates selectable phenotype

Figure 6
Selecting for bacterial mutants favors genes for increased mutation rates. If bacteria in
a population are heterogeneous for mutation rates, selectable mutants will be more
frequent among the more mutable cells. Selecting cells with mutant phenotypes will
enrich the population in cells with higher mutation rates. Enrichment may continue
for several rounds of selection.

because any cells in the original population with a higher mutation rate will have greater probabilities of acquiring the desired mutations. This effect is so strong that the standard way to obtain strains of bacteria and viruses with elevated mutation rates is to select either simultaneously or sequentially for individuals that have several of the desired class of mutations (Cox, 1976). The more mutations selected, the greater is the probability of obtaining organisms with increased mutability.

The logic of this effect extends to all types of evolutionary directors, not just to mutator systems. The populations most likely to evolve in a particular way are those with the most effective evolutionary directors favoring that form of change instead of an alternative. Thus, as a species evolves through the participation of evolutionary directors, that evolution can strength those directors. The species thereby becomes more likely to evolve by related sorts of change in the future. Such patterns and trends in evolution become self-reinforcing.

As a result, species become specialized to evolve in certain ways just as they become specialized to carry out all other biological functions in one way rather than another. Such specialization is apparent in phylogeny. Taxons of organisms tend to adapt, radiate, and speciate in characteristic patterns. This is a premise that underlies conventional taxonomy.

This implies that evolution is more than just adaptation to the current environment. More important, it is a progressive process that teaches the species how to evolve. A gene pool learns by adapting. It learns the strategies and directions that were successful in the past. Those species and lineages that develop inappropriate or ineffective strategies are eliminated, while the surviving ones develop evolutionary directors that bias them to continue their successful modes of adaptation. For other recent discussions on the evolution of evolutionary capacity see Krassilov (1980), Layzer (1980), Conrad and Volkenstein (1981), and Katz (1983).

Sensory Evolution

To the extent that present demands resemble those of the past, the strategy of evolving according to patterns that were successful in the past is sound. It is better than responding without direction to every random gust of selection pressure. However, a species does even better to direct its evolution in accordance with the current state of the environment and organism than the past. For *sensory evolution*, an or-

Darwin Eukarya

kingdom Animalia

Chordata

Vertebrata

Mammalia

Primates

Hominidae

INDUCIBLE EVOLUTION BY THE SOS SIGNAL SYSTEM

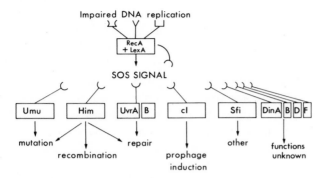

Figure 7

The inducible SOS system of *E. coli*. Boxes represent operons controlled by repressors indicated by squiggles. When chromosomal replication is blocked, the resulting biochemical perturbations induce the RecA operon. Abundant RecA protein serves as an SOS signal to derepress operons with RecA-sensitive repressors. These operons code for DNA repair, mutation and recombination, and various other adaptive and evolutionary functions.

ganism detects relevant information about its environs and influences the way that it evolves accordingly. This capacity might be dismissed as an impossible ability for a species to evolve had we not found mechanisms for it. Actually, two sorts of directors for sensory evolution have been discovered in bacteria. One is exemplified by the repressor for Tn917. This is a sensor, evolved by recruitment, that causes relevant environmental conditions to induce evolutionary activities in particular complex genes. A variety of complex genetic elements with evolutionary functions have their own individual sensory systems (Campbell, 1982).

The other sort of evolutionary sensor is a generalized informational system that evaluates the status of the organism as a whole and broadcasts its assessment to the genome at large. Figure 7 diagrams the *RecA* system of the bacterium *Escherichia coli* with this function (Little and Mount, 1982). This system monitors directly or indirectly several biochemical parameters related to DNA metabolism. When it detects an impairment of chromosomal replication, the *RecA* gene triggers a so-called SOS signal. Various operons have repressors to listen for this signal with *RecA*-sensitive repressors. Some code for DNA repair pathways that might correct the problem that activated them. Their enzymes also promote mutation and generalized and specific recombination to help the organism evolve its way out of difficulty. Other SOS-dependent operons provide a variety of other evolutionary and adaptive functions (Miller, Kirk, and Echols, 1981).

Echols (1981), a prominent molecular geneticist, has coined the term *inducible evolution* for SOS-activated mutation and recombination in *E. coli* and has called attention to its counterparts in higher organisms. Sensory systems to convey information to genes abound in higher organisms, and some of these are known to control mutation, recombination, and the enzymatic alteration of complex genes in somatic cells (Campbell, 1983). It has been proposed that the more elaborate of these sensory systems, notably the immune (Steele, 1979) and endocrine systems (Campbell and Zimmermann, 1982), have been recruited for sensory evolution. It is even possible that the entire morphology of an animal can serve as a sensory device for detecting developmental stress during embryogenesis and regulating the exposure of developmental genes to selection accordingly (Waddington, 1957, and Katz, 1983).

The crucial question for evolutionists now is not whether sensory evolution occurs, but how extensively prokaryotes and eukaryotes have recruited their sophisticated adaptive sensory systems to aid their evolution.

Projective Evolution

Despite its extraordinary potential, sensory evolution is not the ultimate level of control that a species might acquire over its own evolution. The real significance of an adaptation is how adequately it functions after it evolves. The capacity to evolve according to what will occur in the future instead of what the environment is like at present would be a landmark achievement of the highest forms of life.

There have been a variety of earlier proposals of goal-directed evolution, but they had to be vitalistic because biology was too poorly understood to suggest explicit mechanisms. Therefore they were dismissed as nonscientific. Today it has become possible to identify specific genetic and physiological structures that could allow a species to evolve projectively, and to look for evolution that such structures might have directed. In particular, we now know that higher organisms have several adaptations that reasonably could be recruited to serve projective evolutionary functions. It is plausible that mechanisms for *projective evolution* have evolved. An adaptation that can be recruited to a projective evolutionary driver is expected to have the following attributes: (1) access to multiple sources of information about self and the environment; (2) ability to analyze this information to determine which physiological responses will be adaptive; (3) access to genes; (4) access to the germ-line or reproductive system; (5) ability to induce anticipatory physiological adaptations.

These attributes would seem prerequisite for recruitment as a projective evolutionary driver. Both the endocrine and nervous systems of higher animals have all of these features. The endocrines, now more appropriately called cybernins, form an integrated, analytical system of hundreds or possibly thousands of types of signaling molecules controlling every aspect of somatic physiology, development, and reproduction. They allow the organism to adapt physiologically both by responding to and by anticipating stress (like the surge of adrenaline to prepare the organism for imminent fight or flight). I shall describe elsewhere the emerging picture of the endocrine system and how it could act in a projective evolutionary fashion.

The nervous system has even more obvious anticipatory functions. Moreover, as we continue to chart the molecular biology of eukaryotes, we probably will find anticipatory functions in still other control systems. For example, the network of hnRNA molecules in the nucleus (Davidson and Britten, 1979) could conceivably have such analytical capacity, although its function is speculative today.

Any anticipatory capacity of an adaptive control system must be acknowledged to be potentially available for recruitment to projective evolutionary drivers. It is my opinion that the most advanced animals have realized mechanisms for evolving projectively, and that their most significant achievements have been directed by identifiable projective evolutionary drivers.

Domestication as Projective Evolution

A patent example of projective evolution is the development of dwarf strains of rice and wheat through artificial selection. These strains are responsible for the green revolution in tropical agriculture. They were created to grow optimally in highly fertilized tropical soil—an environment that did not come into being until after the strains were produced. Tropical farmers did not add high levels of nitrogen to their land until the highly fertilizer-responsive strains had been evolved. In this case, evolution of the plants caused the suitable environment to develop instead of the other way around.

This example is obviously a very special case, but such must be expected for projective evolution. Its specialness does not detract from its importance. Domestication and artificial selection under man's guidance is overwhelmingly the most significant, rapid, and creative evolution occurring in the biosphere today. Possibly this will remain so for the rest of the existence of life. An adequate theory of evolution must encompass this most recent development of evolution instead of dismissing it. Artificial selection is more than an analogy for natural selection; it is an important development of the latter, a fact obscured only by too narrow and passive a conception of the evolutionary process. There is no justification for excluding domestication by man from evolution while accepting comparable changes that other species have wrought on one another, such as the changes in fungi by the farming practices of ants, or in the fruit characteristics of plants by the predilections and tastes of animals that pollinate them and disperse their seeds. A central argument of Darwin in *The Origin of Species* is that the changes in domestic species are evolutionary. The only seemingly unique specialty of man's domestications is that they are preconceived; but this must be true for any projective evolution.

Of course, domestication does involve the extra complexity (or perhaps simplicity) of one species acting upon another. The dwarf wheat was not the species doing the projection. This point is of secondary importance because coevolution of various sorts is an integral part of the evolutionary process. Also, the same process of domestication logically could occur with just a single species acting upon itself, and probably has. The obvious species to have carried out such evolution is our own. I suggest that man has evolved his human characteristics by a projective process of autodomestication. He evolved himself by practicing the same sorts of manipulations of his breeding system as he also practices on his animals (Zeuer, 1963).

I have been impressed, from admittedly limited reading, with how similarly primitive man seems to treat his wives and offspring on the one hand and his domestic animals on the other (Scott and Fuller, 1965). In fact, some primitive societies still have a standard conversion factor for the worth in pigs of a bride. One can only expect that treatments that led to domestication of animals would produce directly corresponding effects when practiced on the human species.

The origin of man was a unique evolutionary event in both its rapidity and its extent of development. The only obvious novel factor to account for the evolution of the hominid line at this particular time was man himself. As an evolutionary agent, evolving man had opportunities to tamper with his reproduction in ways never before possible. In fact, anthropologists have formally proposed that man's emergence was precipitated by the development of a breeding system that transferred an overwhelming influence over survival to the hands of members of his own species (Lovejoy, 1980). Deliberate parental breeding and social behavior replaced the external environment as the main determinants of fitness of phenotypic traits. Conscious behavior became the driver for primitive man's evolution.

To understand how this driver may have operated, we have the examples of its use in the domestication of other species. Domestication was a complex process that historically proceeded through a series of incremental stages. Darwin (1859) was not content with just the term 'artificial selection'. Instead he recognized several processes that succeeded one another in domestication. Prehistoric man first practiced 'unconscious selection' and only later 'systematic' and 'methodical' selection. Each of these processes presumably was guided, in part, by cultural representations of ideals (Darwin, 1859). Zeuer (1963) has extended Darwin's analysis by recognizing a series of additional stages at the beginning of the process of domestication. He also makes the extremely interesting observation that a variety of social animals besides man (especially among insects) also practice these initial steps of domestication on other species. Possibly some of them also have taken the first steps toward autodomestication as well. Thus autodomestication is not necessarily an anthropomorphic concept.

Man's self-evolution presumably proceeded through a progressive series of steps corresponding to those of domestication. Of course, autodomestication had the extra conceptual complexity that man's emerging mental capacities had to drive their own evolution. The species had to pull itself up by its own evolutionary bootstraps. This probably accounts for the far longer time that it took for man to evolve himself than his recent domesticates.

Primordial man's first manipulations of his reproduction must have been rudimentary, just sufficient to develop a slightly more incisive ability to tamper further. They probably corresponded to a specialized form of sexual or social selection. Many animal societies have a hierarchical organization with social position governing opportunities to reproduce and access to resources. In some primate societies the social status of the individual depends upon his self-image. The individual evaluates himself relative to others and assumes a commensurate status. This process depends upon a system of standards or ideals or values appreciated by the animals of the species. It should also favor the ability of individuals to formulate and to communicate these social standards to others in the population (by allowing the better communicators to promulgate ideals that favor themselves).

Finally, the definitive form for projective evolution would be a system which defines explicit future-related goals and ideals and directs evolution into realizing them. For human autodomestication these projective representations likely were cultural ones, just as Darwin suggested that they were for early and late stages in animal domestication. One cannot help but be impressed with the influence of cultural representation on sexual and reproductive practices, on ideals, and on the indefinite future and past. Cultures have an almost infinite variety of ways for biasing evolution in accordance with these ideals. The concept of the human psyche as fundamentally an evolutionary driver instead of an adaptation for survival poses a profound new dimension for interpreting psychology, sociology, history, and the paradoxical philosophical issues which are so baffling to man as a Darwinian product.

Domestication of plants and animals has probably progressed to a stage beyond that ever applied to humans. We now rationally use the scientific knowledge of genetics and organismic development to achieve exact, predetermined, quantitative goals in agriculture. It is also significant that current agricultural practices are in no way ultimate ones. On the contrary, agriculture is on the brink of a revolution. Through genetic engineering, evolution soon will proceed by the introduction of foreign genes, the chemical modification of extracted genes in the test tube, and even the writing of wholly artificial genes with the aid of computerized DNA-synthesizing machines.

Life by no means has reached the end of evolution viewed as the process of increasing the capacity of the organism to evolve. Current man must also be viewed as only an intermediate in evolution, and the ultimate meaning of his current traits must be their implications for catapulting his higher evolution in the future.

Teleology and Causality

The study of evolution is expanding inexorably to include structure-function interpretation. Accordingly it becomes increasingly evident that organisms evolve special structures to promote their capacities to evolve, and that these structures enormously expand the scope of the evolutionary process. Nevertheless, function is fundamentally a teleological concept, especially when applied to the evolutionary process. Most molecular biologists still skirt this problem by using alternative terms, such as *evolutionary role, evolutionary importance,* and *evolutionary implications* for *evolutionary function.* However, the roles that specially evolved structures play in evolution are so similar—conceptually and even mechanistically—to their roles in adaptation that it is just philosophical prejudice to admit the stature of function to the one and not the other.

For example, the repression system of Tn917 has almost identical effects on the drug-resistant phenotype and on transposability, making the manifestations of both contingent upon the presence of erythromycin. It is unjustifiably arbitrary to accept repression of one set of coding sequences of the transposon as a function of the repressor and that of the other as only a role. Similarly, the function of an enzyme is almost universally identified with the chemical changes that it was evolved to catalyze. It is untenable to single out those gene-processing enzymes whose significance to the species is to enhance its evolution through gene transposition, conversion, elimination, duplication, mutagenesis, and so on as being without functions.

Thus evolutionary functions present an uncomfortable dilemma. As von Brücke reputedly put it, "Teleology is a lady without whom no biologist can live—but he is ashamed to be seen with her in public" (Davies, 1961). Some particularly astute biologists have managed to separate adaptive functionalism from finalism through the concept of teleonomy as a substitute for teleology. Teleonomy holds that the function of a biological structure, whether tooth, nest, lymphocyte or hemoglobin molecule, is not what the structure will usefully do for the organism but the effect that the homologous structure in the ancestors had on survival in past generations. Teleonomic function thereby refers to past effects instead of present purpose. This is a valuable point.

Simple evolutionary function stays within the bounds of teleonomy, even though it implies future change from generation to generation instead of strict repetition. It still emerges from teleonomic selection by consequence. However, as the functions of evolutionary drivers advance from programmed to sensory and finally to projective evolution, they decisively transcend the confines of teleonomy. Scientifically practiced domesticating selection, for example, is patently goal-driven. Either we must devise a more sophisticated sequel to teleonomy for this climax of evolution, or we must abandon the strategy of explaining away finalism in biology and instead work out how it can genuinely occur. The latter approach is the exciting one because it seeks to broaden the foundations for scientific understanding rather than defending their narrowness. I would like to explore this topic by reflecting on the dependence of causality and organization.

The Emergence of Causality from Organization

Scientific explanation rests upon the axiom that causality operates in only one temporal direction. Cause precedes its effects. Obviously, this principle corresponds to the behavior of systems for which science has been successful—classical physics, chemistry, geology, astronomy, and biochemistry. However, some systems have proved intractable to such analysis. These are complex ones with an aura of the teleological in their behavior. The degree to which the so-called social sciences, human (and animal) behavior, and even biological evolution resist conventional reductionistic interpretation suggests that complexly organized systems behave in ways denied by the received assumptions of science. I propose that their fundamental novelty, as suggested by their seeming end-directedness, is to exceed our axiom about causality. When matter is appropriately organized, it becomes sensitive to causes arising from the future instead of just the past.

The idea of reverse causality is not new to philosophy (Mackie, 1966; Waterlow, 1974; and Sanford, 1976). However, these philosophical discussions tend to be at such a metaphysical level that they have little relevance for explanatory science. Most scientific considerations have been aimed at discounting the concept instead of positively exploring its validity and operation. Let me suggest a scientifically acceptable framework for how the future can act causally on the present. This will include the reasons which convince me that (1) the future does play causally on sufficiently organized matter; (2) acceptance of future causality is compatible with objective scientific interpretation; and (3) discovering the physical mechanisms for future causality and their manifestations is essential for understanding and controlling our world.

Contrary to its superficial appearance, causal behavior is not an inherent property of matter as such. Instead, causality emerges only from organization of matter. Poorly organized material admits only rudimentary forms of causality. As matter becomes progressively more

Table 1
Causality and Organization

Type of Organizations	Emergent Causal Property	Format for Causal Analysis
None (elementary particles only)	Acausal	Quantum mechanics
Mechanical objects	Deterministic cause and effect	Newtonian mechanics
Negentropy	Unidirectional cause and effect	Thermodynamics
Information		Cybernetics
Information about self	Recursive causes are their effects	Self-reference
Information about future self	Future causality	Future self-reference

organized it sequentially acquires qualitatively new capacities for causal interactions. In table 1 a hierarchical series of qualitatively distinct types of material organization exhibits new causal qualities which require different principles for causal analysis. Note that each level of organization manifests causal qualities absent at lower levels.

Quantum mechanical organization

Matter in its most elementary state behaves acausally in accordance with quantum mechanical principles. Quantum mechanics is explicitly statistical instead of deterministically causal. As a familiar example, the timing of the spontaneous decay of radioactive atomic nuclei is determined only as statistical probability of occurrence. No pathway of hidden variables exists to cause an individual particle to undergo a permissible transformation at a particular time. Unorganized elementary matter is without capacity for necessary and sufficient conditions to ensure predictable effects.

Probabilistic quantum mechanics is so alien to intuition that physicists still disagree on how to interpret its implications. However, there is no dispute that our familiar notion of local deterministic causality is only a secondary characteristic of the material universe. It appears only when matter is organized into stable composite structures.

Mechanical organization

The simplest sort of material organization is the aggregation of elementary particles into objects. These are entities that occupy discrete spaces distinct from their surroundings. Objects interact causally with one another in a mechanical way. For example, moving balls on a billiard table will cause each other to change their speed and direction of movement. These causal interactions between mechanical objects are mediated by measurable forces, and Newtonian mechanics provides an explicit, mathematically precise, and deterministic formulation for their effects: $F = dmv/dt$. The present state of a mechanical object can be rationalized as the effect of the total history of causal force imposed upon it by other mechanical objects. Of course, acausal quantum mechanical behavior still underlies these dynamics, but its statistical properties in the aggregate correspond to essentially causal interactions on a macroscale.

Mechanical organization supports causality, but only in a weak form. Although one can identify corresponding causes and effects, the distinction between them is arbitrary. By convention, the cause precedes its effect. Yet Newtonian physics places no arrow on the direction of time. A movie of two balls bouncing along on an idealized, frictionless billiard table can be shown backwards or forwards indiscriminately. When shown backwards, the effects of collisions assume the roles of causes and obey exactly the same mathematical laws as they did the other way around. Newtonian cause and effect are qualitatively indistinguishable except arbitrarily by temporal order.

Thermodynamic organization

When material is given a higher level of organization, effects do become quantitatively differentiated from their causes. The extra organization can be incorporated as internal thermodynamic order or negentropy. A mechanical system whose internal components are not at statistical equilibrium assumes behavioral properties described by the second law of thermodynamics. Change increases overall entropy. Thermodynamic causes dissipate order and energy so that improbable or organized causal states invariably lead to less improbable and less ordered effected states. One cannot surreptitiously reverse a movie of the breaking shot on a pool table. The coalescence of scattered balls into a perfect triangle emitting a cue ball with all of the original energy and momentum does not correspond to reality. To be sure, reversible causal processes can occur briefly or locally in a thermodynamic system, but their effects will be swamped out globally and eventually by the cumulative dissipative effects of irreversible entropy-generating causes.

Cybernetic organization

Matter can possess still higher degrees of organization in the form of internal information. Information is an elaborate relative of entropy, and hence of organization. It can be static, like the indelible words written in a book, or more interestingly, it can be dynamic, as, for example, informational signals being processed in a radio or computer circuit.

Dynamic processes involving information can exhibit causal properties not encountered in less organized systems. A revealing example is a public address system. When such a system operates properly, we can identify the words spoken into the microphone as the cause of the words issuing from the loudspeaker. The one implies and determines the other. We can even follow the course of the signal as a series of mechanistic processes along the circuit. Sound waves propagated through the air cause a coil wire to move through a magnetic field in the microphone. This movement induces an electric current in the coil. At the amplifier the signal current causes an amplified current to be sent to the speaker, and so forth. Each of these processes conforms to simple cause-and-effect principles.

The PA system can manifest more complex causal properties if its components are globally arranged into a feedback loop. Inadvertently placing the microphone too close to the speaker still allows spoken words to be amplified, but also generates the familiar feedback screech that eventually drowns them out. The screech develops when sound emanating from the loudspeaker reenters the microphone. Each time the sound feeds back through another amplification cycle it builds in intensity until the amplifier is saturated.

A cause-and-effect analysis of this recursive process leads to a remarkable conclusion. Effects cannot be separated from their causes. In fact the cause becomes part of its effect. This peculiarity is especially apparent for a feedback screech initiated by a very sharp click. Initially the signal has a short duration. As it progresses through the PA system the signal occupies only a limited segment of the circuit. However, the signal inevitably dissipates, occupying a longer segment of time and a greater distance along the circuit. Eventually the leading edge of the signal overtakes its tail when the signal lasts longer than the transit time through the circuit. Now the leading edge is part of what we identified as an effect and the tail as the cause of that effect, having been around the circuit one less time. Yet where the two merge it is meaningless to distinguish between them. Eventually the entire causal signal blends with its effect. Later it also merges with the effect of that effect, and so on. The final signal includes the historical series of its prior causal states expressed simultaneously as components of their effect.

A PA system also systematically modifies the signal with each cycle through losses and amplification. This causes the signal to evolve toward a stable form determined by the circuitry. The reader may recall from experience that feedback interferences in PA systems often start out at a moderate pitch and progress to high-frequency screeches.

As one might expect from these complexities, simple cause-and-effect analysis fails to rationalize the outcome of recursive cybernetic processes. In particular, the causal threads that I discussed earlier as originating from mechanical and thermodynamic levels of organization are historical. One can meaningfully understand the characteristics of such systems as the end effects of a historical sequence of preceding causal events. This fails for a feedback signal. All of the characteristics of the final screech—its frequency spectrum, intensity, stability, and so forth—are determined entirely by the organizational characteristics of the feedback circuitry. From the particular circuit components of a misaligned PA system, one can predict exactly what its screech will be like. In contrast, the final form is independent of the frequency, information content, and duration of the sound waves that originally seeded the circuit. One can predict nothing from these details. Also, the final state is immune to perturbations incurred during its formation. Briefly distorting a developing signal in a simple feedback loop leaves no mark on the final state.

The companion inadequacy of historical analysis is the impossibility of deducing from the final state of a recursive signal what originally set it off or how many cycles it has gone through. In fact a sufficiently sensitive feedback circuit does not even need initiating signals to activate it. The circuit will spawn a signal spontaneously through inevitable thermal noise or even fundamental quantum instability, given enough time. Appropriate cybernetic organization of itself can imply that genesis of a self-reinforcing signal within it. Thus organization that allows effects to become synonymous with their cause also permits these effects to evolve spontaneously, as though they act as their own cause.

This sort of behavior is difficult to accommodate in traditional cause-and-effect analysis patterned after the dynamics of Newtonian or thermodynamic systems. Yet it cannot be ignored. Spontaneity is the most notable and insidious characteristic of fulminating positive feedback signals. Designers of electronic circuitry, computer programs, and even auditoriums must assiduously watch out for internal organization that inadvertently forms uncontrolled recursive circuits. Uncontrolled positive feedback loops are eliminated, categorically, regardless of the likelihood or the sort of signals which might trigger them off. These units of organization are respectfully treated as sources of problems, that is, as causes of signals that develop spontaneously in them.

Spontaneity of recursive signals also has special implications for analyzing biological phenomena ranging from embryonic development (Kato et al., 1981; Canalis, et al. 1980) to cancer (Kaplan et al., 1982; Todaro et al., 1977; Liu et al., 1979) to the etiology of schizophrenia (Nicoll and Jahr, 1982). For example, a main concern of experimental embryology has been to identify the chemical inducers that cause various developmental processes to occur in a precisely set sequence in time and space during embryogenesis. The troubling finding has been that a large, miscellaneous array of chemical conditions and treatments, some very artificial, act perfectly well as inducers for many elaborate developmental pathways. It is now suspected that developmental changes are programmed in a very different manner from controlling the inducers of change. Tissues may be internally organized so that when a biochemical perturbation occurs it will cycle recursively and evolve itself and the tissues into a specified final form. The nature of the inducer that triggers the recursive circuit is irrelevant. Indeed, the organism may not even genetically specify a particular inducer, since the organism can count on noise inevitably being available to induce any self-perpetuating recursive process. The classical concept of equating cause with initiator and hence of assuming that developmental processes are orchestrated by coordinating these primary causes may be completely misplaced. It chases after the noise in the system instead of the effective cause. The same principle also operates on the process of evolution (Kirkpatrick, 1982).

Of course a signal in a feedback system does not violate historical cause and effect locally. One can successfully trace a small segment of a signal through one or two cycles of a recursive circuit, such as a misaligned PA system in a historical cause-and-effect fashion. It is just that such analysis is inadequate for rationalizing the behavior of the system as a whole. Eventuating global dynamics can be understood only as effects of the contemporary cybernetic organization of the system. It is the way that the system is organized that causes the signals within it to acquire particular, predictable dynamic properties. If one wishes to understand or to control systems with cybernetic levels of organization, he must be concerned with these new emergent causal qualities.

Causality in ultra-organized systems

Can one expect still further causal qualities to emerge if matter is organized to even higher degrees? The progressions shown in Table 1 encourage such a belief, and observation supports this conjecture. For example, the brain is the most highly organized form of matter known (McGeer, Eccles, and McGeer, 1978), and it exhibits unprecedented causal properties such as free will, intention, anticipation, choice, and valuation. These properties are so foreign to causality as manifested by less organized systems that they generally are either simply dismissed by science as illusory, as in Skinnerian behaviorism, or totally dissociated from material causality, as in Cartesian dualism. Sperry (1969) has proposed a scientific interpretation: The brain truly does manifest the novel causal properties that it appears to have. These causal properties emerge in toto from the brain's supreme organization.

Mentality, then, is an enriched mode of causality that emerges from the special organizational state of the brain, and the mind can be identified with the organization that gives rise to mental cause. The mind as organization is causally responsible for the behavior of the brain in the same sense that negentropy can directionally drive chemical and physical reactions, and that recursive cybernetic organization can create internal information signals. A special value of viewing mentality as an attribute of causality is that the only publicly accessible aspect of mentality is its causal influence on behavior. It thus circumvents the metaphysical silliness of mentality as an inherent cryptic quality of matter infusing even inanimate mechanical objects. Mentality would be categorically restricted to matter with the prerequisite level of organization. It would emerge de novo from the evolution of a previously unprecedented type of material organization (Dobzhansky, 1967) instead of having to be immanent in matter itself, and only brought into manifestation by the evolution of the brain. Emergence of the causal mind as organization in the brain makes it amenable to biological structure-function analysis. One approach to understanding mind would be to deduce the causal properties that understandably might emerge from very high degrees of organization, an attractive inquiry even for the non-neurophysiologist.

Such considerations suggest that scientists will bring to light further new, richer attributes of causality as they successfully study ever more complexly organized systems. Even the capacity of causality to operate from the future to the present loses its utter absurdity as an emergent property of the right sort of ultracomplex organization. The scientific question for future causality is not whether it is possible, but the mechanistic one as to what sorts of organizational features could sensitize the behavior of matter to its future form. Such organization would have to produce relationships that transcend current axioms of scientific logic, and make the future operationally available somehow to the present. Significantly, appropriate organization can meet these two demands.

Self-referent organization

A higher level of internal organization than simple information is self-referent information. This is information about the system which embodies it. A record player can house and process any sort of information or noise, but a self-referent record would describe its own form. Self-referent systems are substantially more organized than those discussed above, and can exhibit truly astonishing properties. Gödel discovered that self-reference transcends the bounds of traditional logic. It admits relationships (called "strange loops" by Hofstadter, 1979) that are incompatible with the fundamental tenets of logical analysis. Self-referent statements such as, This statement is false, can be written in algebraic logic, but defy the core axiom that every logical proposition must be either true or false. Gödel showed that formal logic is valid only for systems that do not reach the organizational level or self-reference. Strange loops defy logic, but have analogies with feedback loops.

Future self-reference

A particularly advanced form of self-reference is future self-reference. Future self-referent information describes not its current physical housing, but the future state of the material system that embodies it. A future self-referent system that also has mechanical ability to operate on the physical world around it is capable of extraordinary causal behavior. It can modify the structure of itself and its surroundings under the direction of its internal future description so as to assume the form that it internally describes. This allows a representation of the future to direct activity in the present (Falk, 1981). In effect, future self-reference gives the future access to act causally on the present.

It is of utmost importance to recognize that these directive future self-referent models have a more imperative status than predictions of the future. They are also commands that cause the physical systems under their influence to materialize them in the future. They have the added certainty of, say, the statements of the corrupt bookie that not only predict which horse will win, but also serve as directions to the jockeys to cause them to fix the race, thus ensuring the outcome. Insofar as the future self-referent system can discriminately entertain directive models of the future that are within its capacity to fulfil, those models will eventuate. They will eventuate because and only because they are future self-referently modeled.

Future causality is the relationship of such attainable preconceived future organizational states to the events that bring them into being through preconception. Future causal analysis inquires into the qualities of a future situation that are necessary and sufficient for them to emerge from the behavior induced by future self-reference, and into the roles of these qualities in the self-realizing process. It implies that the future acts causally on the present through the proxy of a descriptive prerepresentation.

This capacity of future self-referent organization represents a major new emergent causal quality. It allows effects to temporally precede their causes. Because the future is still viewed as a metaphysical agency by science, it is essential to distinguish a weak and a strong interpretation of future causality. The strong statement is that the future genuinely can act causally upon the present through future self-reference. The weak statement avoids any metaphysical implication. It asserts only that only future self-referent systems can behave as though the future were acting causally upon them, although this is just mimicry. Having a hypothetical model of the future inside itself can cause a system to behave in a way that resembles that expected if the system truly did have a foreknowledge of the future or if the future actually could reach back causally to determine the behavior of the system in the antecedent present. With this distinction between weak and strong future causality we can rationally explore the characteristics of behavior that future self-referent information can produce unencumbered by the philosophical issue involved. Elsewhere I shall discuss the legitimacy of the strong interpretation as an operating causal principle in the real world. I shall try to show that if one accepts the weak interpretation he cannot rationally fail also to accept its extrapolation to the strong interpretation. However, this issue requires discussion of several other aspects of organization that cannot be covered in the pages available here. Nevertheless, despite the obviously much greater significance of the strong interpretation, let me point out that the weak interpretation is sufficient to rationalize the degree of end-directedness shown by the types of evolution suggested here. Weak future causality is a substantial advance over teleonomy. Future self-referent organization implies overtly teleological behavior. Surely this is a reason to embrace the concept instead of rejecting it as scientifically unacceptable. Apparent end-directedness is the most striking and important characteristic of future self-referent processes. This includes biological evolution, embryonic development, and various activities of people and their cultural-social institutions.

It is true that teleological interpretation is considered both unacceptable and antiscientific for these systems. In fact, nearly every scientist who has written on the general nature of evolution has felt compelled to show how deftly he can skate toward the abyss of teleology without falling in (Davies, 1961). Anyone who recognizes that scientific paradigms are postulates instead of absolute truths must be surprised that evolutionists feel so strongly compelled to rationalize away evolutionary teleology as a mirage, instead of searching for the meaning or significance of its end-directedness. Evolutionists seem more dedicated to defending current philosophical viewpoints from challenge than to extending their understanding. The abuse and philosophical intractability of teleology in evolution theory during the past century is no reason to dismiss the issues behind it today. To understand its end-directed properties is the most important reason for studying the uniquely complex natural process of evolution. I do not hold that we should accept nebulous teleology as an explanation. But we should investigate the mechanisms that allow finalistic behavioral characteristics where they emerge. Involvement of future organization in the dynamics of the present, if it occurs, would be a most important discovery to make.

The poorest possible reason for rejecting projective evolution is that it may imply a substantial alteration of our conception of reality.

Note

This work was carried out while I was a visiting professor of the Genetics Department, Research School of Biological Sciences, Australian National University, under the support of NSF Grant PCM81-20923. I am indebted to the kind hospitality of Barry Rolfe and John Pateman of the Genetics Department and to the many provoking ideas of John Langridge of the CSIRO in Canberra.

References

Anderson, E. S., 1968. The ecology of transferable drug resistance in Enterobactereacae. *Ann. Rev. Microbiol.* 22:131–180.

Campbell, J. H., 1982. Autonomy in evolution. In *Perspectives on Evolution*, R. Milkman, ed. Sunderland, Massachusetts: Sinauer Assoc., pp. 190–200.

Campbell, J. H., 1983. Evolving concepts of multigene families. *Isozymes* 10:401–417.

Campbell, J. H., and E. G. Zimmermann, 1982. Automodulation of genes: A mechanism for persisting effects of drugs and hormones in mammals. *Neurobehav. Tox. Terat.* 4:435–439.

Canalis, E., W. A. Peck, and L. G. Raisz, 1980. Stimulation of DNA and collagen synthesis by autologous growth factor in cultured fetal rats calvaria. *Science* 210:1021–1023.

Chou, J., P. G. Lemaux, M. J. Casadaban, and S. N. Cohen, 1979. Transposition protein of Tn3: identification and characterization of an essential repressor-controlled gene product. *Nature* 282:801–806.

Conrad, M., and M. V. Volkenstein, 1981. Replacability of amino acids' self-facilitation of evolution. *J. Theoret. Biol.* 92:293–299.

Cox, E. C., 1976. Bacterial mutator genes and the control of spontaneous mutation. *Ann. Rev. Genet.* 10:135–156.

Cox, E. C., and C. Yanofsky, 1967. Altered base ratios in DNA of an *Escherichia coli* mutator strain. *Proc. Natl. Acad. Sci.* 58:1895–1902.

Darwin, C. R., 1859. *The Origin of Species.* London: Murray.

Davey, R. B., and D. C. Reanney, 1980. Extrachromosomal genetic elements and the adaptive evolution of bacteria. *Evol. Biol.* 13:113–147.

Davidson, E. H., and R. J. Britten, 1979. Regulation of gene expression: possible role of repetitive sequences. *Science* 204:1052–1059.

Davies, B. D., 1961. The teleonomic significance of biosynthetic control mechanisms, *Cold Spring Harbor Symposia in Quantitative Biology* 26:1–10.

Dobzhansky, T., 1967. *The Biology of Ultimate Concern.* New York: North American Library.

Dover, G., 1982. Molecular drive: a cohesive mode of species evolution. *Nature* 299:111–112.

Echols, H., 1981. SOS functions, cancer and inducible evolution. *Cell* 25:1–2.

Falk, A. E., 1981. Purpose, feedback and evolution. *Phil. Soc.* 48:198–217.

Flanagan, J. G., and T. H. Rabbits, 1982. Arrangement of human immunoglobulin chain constant region genes implies evolutionary duplication of a segment containing γ, E and α genes. *Nature* 300:709–713.

Grindley, N. D., 1983. Transposition of Tn3 and related transposons. *Cell* 32:3–5.

Grobstein, C., 1974. *The Strategy of Life,* 2nd ed. San Francisco: W. H. Freeman & Co.

Hiraizumi, Y., L. Sandler, and J. F. Crow, 1960. Meiotic drive in natural populations of *Drosophila melangoster:* III Population implications of the segregation distortion locus. *Evolution* 14:433–444.

Hofstadter, D., 1979. *Gödel, Escher and Bach: An Eternal Golden Braid.* New York: Basic.

Hood, L., J. H. Campbell, and S. C. R. Elgin, 1975. The organization, expression and evolution of antibody genes and other multigene families. *Ann. Rev. Genet.* 9:305–353.

Hunkapiller, T., H. Huang, L. Hood, and J. H. Campbell, 1982. The impact of modern genetics on evolutionary theory. In *Perspectives on Evolution,* R. Milkman, ed. Sunderland, Massachusetts: Sinauer Assoc., pp. 164–189.

Jacob, F., and J. Monod, 1961. On the regulation of gene activity. *Cold Spring Harbor Symp. Quant. Biol.* 26:193–211.

Kaplan, P. L., M. Anderson, and P. Ozanne, 1982. Transforming growth factor(s) production enables cells to grow in the absence of serum. *Proc. Natl. Acad. Sci.* 79:485–489.

Kato, Y., Y. Nomura, M. Tsuji, M. Kinoshita, H. Ohmae, and F. Suzuki, 1981. Somatomedin-like peptide(s) isolated from fetal barine cartilage. *Proc. Natl. Acad. Sci.* 78:6831–6835.

Katz, M. J., 1983. Ontophyletics: studying evolution beyond the genome. *Perspectives in Biol. and Med.* 26:313–333.

Kirkpatrick M., 1982. Sexual selection and the evolution of female choice. *Evolution* 36:1–12.

Kleckner, N., 1981. Transposable elements in prokaryotes. *Ann. Rev. Genet.* 15:341–404.

Krassilov, V. A., 1980. Directed evolution: a new hypothesis. *Evol. Theory* 4:203–220.

Langridge, J., 1969. Mutations conferring quantitative and qualitative increases in β-galactosidase activity in *Escherichia coli*. *Mol. Gen. Genet.* 105:74–83.

Layzer, D., 1980. Genetic variation and progressive evolution. *Amer. Natur.* 115:809–826.

Little, J. W., and D. W. Mount, 1982. The SOS regulatory system of *Escherichia coli*. *Cell* 29:11–22.

Liu, S. T., C. L. Perry, C. L. Schardl, C. I. Kado, 1979. Agrobacterium Ti plasmid indoleatic acid gene is required for crown gall oncogenesis. *Natl. Acad. Sci.* 79:2812–2816.

Lovejoy, C. O., 1980. The origin of man. *Science* 211:341–350.

Lyttle, T. W., 1977. Experimental population genetics of meiotic drive systems: I pseudo-Y chromosomal drive as a means of eliminating cage populations of *Drosphila melanogoster*. *Genetics* 86:413–445.

Mackie, J. L., 1966. Direction of causation. *Phil. Rev.* 75:41–446.

McClintock, B., 1965. Genetic systems regulating gene expression during development. *Devel. Biol. Suppl.* 1:84–112.

McGeer, P. C., J. C. Eccles, and E. G. McGeer, 1978. *Molecular Neurobiology of the Mammalian Brain*. New York: Plenum Press.

Miller, H., M. Kirk, and H. Echols, 1981. SOS induction and auto-regulation of the himA gene for site specific recombination in *Escherichia coli*. *Proc. Natl. Acad. Sci.* 78:6754–6758.

Monod, J., 1957. Remarks on the mechanism of enzyme induction. In *Units of Biological Structure and Function*. New York: Academic Press, pp. 7–28.

Nicoll, R. A., and C. E. Jahr, 1982. Self-excitation of olfactory bulb neurones. *Nature* 296:441–444.

Pittendrigh, C. S., 1958. In *Behaviour and Evolution*, A. Roe and G. G. Simpson, eds. New Haven: Yale University Press, p. 391.

Sanford, D. H., 1976. The direction of causation and the direction of conditionship. *J. Phil.* 73:193–207.

Scott, J. P., and J. L. Fuller, 1965. *Genetics and the Social Behaviour of the Dog*. Chicago: University of Chicago Press.

Simpson, G. G., 1950. *The Meaning of Evolution*. London: Oxford University Press.

Skinner, B. F., 1981. Selection by consequences. *Science* 213:501–504.

Sperry, R., 1969. A modified concept of consciousness. *Psychol. Rev.* 76:532–536.

Sperry, R. W. 1977. Forebrain commissurotomy and conscious awareness. *J. Med. Phil.* 2:101–126.

Spradling, A. C., and G. M. Rubin, 1981. *Drosophila* genome organization: conserved and dynamic aspects. *Ann. Rev. Genet.* 15:219–264.

Steele, E. J., 1979. *Somatic Selection and Adaptive Evolution: On the Inheritance of Acquired Characters.* Toronto: Williams & Wallace International.

Sueoka, N., 1965. On the evolution of macromolecules. In *Evolving Genes and Proteins*, V. Bryson and H. J. Vogel, eds. New York: Academic Press, pp. 479–494.

Temin, H. M., and W. Engels, 1984. Movable genetic elements and evolution. In *Evolutionary Theory: Paths into the Future*, J. W. Polland, ed. New York: Wiley, pp. 173–201.

Todaro, G. D., J. E. De Larco, L. P. Nissley, and M. M. Rechler, 1977. MSA bind EGF receptors on sarcoma virus transformed cells and human fibrosarcoma cells in culture. *Nature* 267:526–528.

Tomich, P. K., F. Y. An, and D. B. Clewell, 1978. A transposon (Tn917) of *Streptococcus faecalis* that exhibits enhanced transposition during induction of drug resistance. *Cold Spring Harbor Symp. Quant. Biol.* 43:1217–1221.

Van Valen, L. M., 1973. A new evolutionary law. *Evolutionary Theory* 1:1–30.

Waddington, C. H., 1957. *The Strategy of the Genes.* London: Allen and Unwin.

Waterlow, S., 1974. Backward causation and continuity. *Mind* 83:372–387.

Zeuer, F. E., 1963. *History of Domesticated Animals.* New York: Harper & Row.

Self-Organization, Selective Adaptation, and Its Limits:
A New Pattern of Inference in Evolution and Development

Stuart A. Kauffman

Current evolutionary theory is in a state of healthy flux. Our general framework for considering the relation between ontogeny and phylogeny remains dominated by the Darwinian-Mendelian marriage. Within that conceptual marriage, Mendelian segregation, and mutations that are random with respect to fitness, provide mechanisms to maintain the population variance upon which selection acts. Selection is viewed as providing disruptive, directional, or stabilizing forces acting on populations, leading to phenotypic divergence, directional change, or stasis (Simpson, 1943). The early dominant conceptual reliance on selection as the preeminent evolutionary explanatory principle has led to debates about the coherence of a panadaptationist program in which all features of organisms might, in some sense, be maximally adapted (Gould and Lewontin, 1979). The discovery of abundant genetic variance within and between populations has led to debates about the extent to which random drift and fixation versus disruptive, directional, or stabilizing selective forces determine population genetic behavior (Lewontin, 1974; Ewens, 1979). However, the possibility that divergence, directional change, or stasis in the face of arbitrary mutational events might be partially due to the constraining self-organized features of organisms (Rendel, 1967; Waddington, 1975; Bonner, 1981; Rachootin and Thompson, 1981; Alberch, 1982; Wake et al., 1983) has not been incorporated as a truly integral part of contemporary evolutionary theory.

The relative inattention to the contribution of the organism toward its own evolution, that is, to internal factors in evolution, finds one of its conceptual roots in the very success of the Darwinian-Mendelian heritage. This heritage replaced the typological paradigm of the rational morphologists, who sought underlying universal laws of form (Webster and Goodwin, 1982) with an emphasis on population biology and continuity by descent. In turn this shift has led to the curious character of biological universals in our current tradition.

While not actually strictly required by the Darwinian-Mendelian heritage, it is nevertheless true that we have come to regard organisms as more or less accidental accumulations of successful characters, grafted onto one another piecemeal, and once grafted, hard to change. This view resurfaces in our deeply held beliefs about biological universals, which appear to us as historical contingencies—accidental but typically useful properties which are widely shared by virtue of shared descent. Thus the code is universal by virtue of shared descent, vertebrates are tetrapods by shared descent, the vertebrate limb is pentadactyl by shared descent. Yet presumably a different code would have sufficed, and vertebrates might have six limbs, all heptadactyl. Whether due to selection or random drift and fixation, all are widely shared but historically contingent properties. The underpinning for this view derives from the abundant evidence for descent with minor modifications. Massive changes are massively disruptive, monsters are typically hopeless, only minor changes typically survive and pass the selection-drift filter.

It is easy to see that this tradition is at best weakly predictive, hence that phenomena such as widespread morphological stasis despite substantial genetic variance find no ready account in contemporary theory (Gould and Eldredge, 1977; Schopf, 1980; Wake et al., 1983; Charlesworth et al., 1982). Gradual change and accumulation of properties thereafter fixed is easily understood. Beyond this, and appeal to design principles and maximal adaptation, we have few conceptual tools to predict the features of organisms. One way to underline our current ignorance is to ask, if evolution were to recur from the Precambrian when early eukaryotic cells had already been formed, what organisms in one or two billion years might be like. And, if the experiment were repeated myriads of times, what properties of organisms would arise repeatedly, what properties would be rare, which properties were easy for evolution to happen upon, which were hard? A central failure of our current thinking about evolution is that it has not led us to pose such questions, although the answers might in fact yield deep insight into the expected character of organisms.

Generic Properties of Self-Organizing Systems as Ahistorical Universals

Imagine, then, that evolution had recurred a myriad of times, allowing a catalogue of common and rare properties. How would we try to think about the common properties? Presumably they are common by virtue of two quite different mechanisms. The common properties might reflect recurrent selection of the same useful features. Conversely, the common features might reflect properties of organisms so easily found in evolution as to be essentially unavoidable. Restated, perhaps there are properties of organisms which recur, not by virtue of selection, but by virtue of being inherent properties of the building materials. Such properties would constitute ahistorical universals. It is this theme I wish to explore, for I want to suggest that many aspects of contemporary organisms may reflect inherent, generic properties of self-organizing systems, and the limited capacity of selection to cause deviation from those inherent properties. Those generic properties, then, will function as something like normal forms. The core of the pattern of inference I want to consider is this: Understand the generic structural, organizational, and dynamical properties of the systems used in generating organisms. Understand the extent to which selection can change those inherent properties. If selection can only cause modest deviations from the inherent properties, then organisms would be expected to exhibit those inherent features. This new typological pattern of inference might be useless but for these considerations: In a variety of ways, some discussed below, very powerful self-organizing properties in organisms, constituting developmental constraints, can be expected and can begin to be sketched. The limitations in the capacity of selection to modify such properties are open to investigation. Experimental implications are already available and should increase. I would note that this pattern of inference is almost completely absent from our current theories.

Two Preliminary Cases

Evolution has not recurred ab initio; we can, however, ask whether there may be generic ahistorical features of organisms. We can do this by considering whether there are widely shared properties among distant phyla which seem highly unlikely to reflect either selection or shared descent. I briefly mention two.

Rhythmic phenomena are found in organisms at many levels. Among these are circadian rhythms in animals and plants, neural periodicities, firefly flash rhythms, yeast glycolytic oscillations, and cardiac rhythms (Winfree, 1980). A common experimental technique to investigate all these rhythms is to apply a perturbation such as a temperature shock, or light flash, then study the alteration in the phase of the rhythmic phenomenon. Application of the same perturbation at all phases results in a phase-resetting curve mapping new phase as a function of old phase. In his recent book, Winfree (1980) has argued that such phase-resetting phenomena have essentially universal propeties which recur over and over, in these myriad circumstances. It is conceivable that the common properties recur on such a wide range of levels and phyla due to recurrent selection for similar adapted phase-resetting behavior. Winfree's own argument seems more plausible. The nearly universal properties are elegantly understood as generic topological properties of continuous oscillatory systems. Build a biochemical oscillator; it is very likely to exhibit a restricted family of phase-resetting behavior, not due to selection, but because those features are generic to such oscillations.

In all higher metazoan and metaphyton phyla, each cell type during ontogeny differentiates directly into rather few other neighboring cell types, along branching developmental pathways. It is logically conceivable, and might be highly advantageous, for a single cell type to proliferate, then differentiate directly into all the very many other cell types found in the adult, after which those cell types would rearrange themselves into the adult morphology. Indeed, dissociated sponges are capable of this rearrangement dance (Humphreys, 1963; Moscana, 1963), yet normal sponge ontogeny follows typical branching developmental pathways (Willmer, 1960). It might be the case that this feature of virtually all metazoan and metaphyton phyla reflects recurrent selection for a common property of high adaptive advantage. I have described in detail elsewhere (Kauffman, 1969, 1974, 1982) and shall suggest below a contrary thesis: Due to the self-organizing dynamical properties of complex genomic regulatory systems, any cell type can generically have but few accessible neighboring cell types. This is no trivial constraint. It implies that higher organisms must rely upon branching pathways of development, with a restricted number of branches at each branch point. If this view has merit, ontogenies for the past billion years have been largely constrained to utilize this pattern, not because it is adaptive, but because it is essentially unavoidable.

I have chosen these two preliminary examples because the properties that appear to be generic are not trivial, but are already important organizing features of higher plants and animals. Further, one feels impelled here to admit that a selective explanation is suspect, and that the properties in question might possibly reflect self-organizing features. Yet selection surely operates, and the fundamental problem I want to discuss is how to begin to think about the relation between generic self-organizing features of organisms, the action of selection which may attempt to modify those properties, and the balance struck by selection and the restoring forces of random mutation which tend to drive the selected system back toward those properties which are generic to it; hence the limits of selection.

The "Wiring Diagram" of the Genomic Regulatory System

In a sense the genomic system in a cell is loosely analogous to a computer. A mammalian cell has enough DNA to encode on the order of 1,000,000 average sized proteins (Bishop, 1974). Current estimates in echinoderms (Hough et al., 1975; Kleene and Humphrey, 1977), and vertebrates (Bantle and Hahn, 1976; Hastie and Bishop, 1976; Chikaraishi et al., 1978) suggest that, in fact, from about 50,000 to 100,000 distinct transcripts are actually found in the heterogenous nuclear RNA, and a subset of these is processed to the cytoplasm as mature message. In eukaryotes, as in prokaryotes, the genomic system includes structural genes which code for proteins, and both *cis*-acting and *trans*-acting regulatory genes which control the expression of structural and other regulatory genes. The modes of action and varieties of regulatory genes are still only partially known. However, it is already clear on genetic and direct molecular evidence that cis-acting sites regulate the transcriptional activity of structural genes in some local domain on the same chromosome. Trans-acting genes typically lie at distant positions on the same or other chromosomes, and presumably exert their influence on a given structural gene indirectly, via diffusible products which interact with cis-acting sites near the structural gene (Green, 1980; Britten and Davidson, 1971; Abraham and Doane, 1978; Paigen, 1979; Davidson and Britten, 1979; Dickenson, 1980a; Kurtz, 1981). Regulation of gene expression occurs at the transcriptional level, in the processing of heterogenous nuclear RNA to the cytoplasm as mature messenger RNA, in the translation to protein, and in post-translational modifications of the protein which modulate its activity (Brown, 1981). The loose analogy of this complex system to a computer lies in the fact that the 100,000 or so different genes and their products form a network whose components regulate one another's activities. Thus a cell is a dynamical system whose complete accounting would require at least specification of all the particular regulatory interactions, that is, a specification of the wiring diagram showing which components affect which components, and the local rule for each component describing the

CHROMOSOME 1   CI TI SI—C2 T2 S2—C3 T3 S3—C4 T4 S4—

CHROMOSOME 2   C5 T5 S5—C6 T6 S6—C7 T7 S7—C8 T8 S8—

CHROMOSOME 3   C9 T9 S9—CIO TIO SIO—CII TII SII—CI2 TI2 SI2—

CHROMOSOME 4   CI3 TI3 SI3—CI4 TI4 SI4—CI5 TI5 SI5—CI6 TI6 SI6—

Figure 1a
Hypothetical set of 4 haploid chromosomes with 16 kinds of cis-acting (*C*1, *C*2, . . .),
trans-acting (T1, T2, . . .) and blank genes arranged in sets of 4 *Cx*, *Tx*, *Sx*.

behavior of that component based on the behaviors of the components
directly affecting it.

To further this discussion, I wish to idealize the very complex system
presumed to exist in eukaryotic cells by which one gene influences the
expression of another gene. I shall idealize the genomic regulatory
system by imagining that the genome consists of structural genes whose
transcription is regulated by nearby cis-acting genes, that each cis-
acting gene regulates all transcribable genes in some local domain that
is demarcated in some way, that trans-acting genes lie in the domain
of cis-acting genes, which regulate the expression of the adjacent trans-
acting genes, while the trans-acting genes themselves produce products
targeted to act on specific "matching" cis-acting sites anywhere in the
chromosome set.

This crude, but current picture of genetic regulation in eukaryotes is
captured in figure 1a.

Here I have assumed that the genome has 4 kinds of genetic elements,
cis-acting (*Cx*), trans-acting (*Tx*), structural (*Sx*), and empty (−) domain-
demarcating elements. For concreteness I have assumed a haploid organism with 4 chromosomes, having these kinds of genes dispersed
in the chromosome set. Any cis-acting site is assumed to act in polar
fashion on all trans-acting and structural genes in a domain extending
to its right to the first blank locus. Each indexed trans-acting gene, *Tx*,
is assumed to regulate all copies of the corresponding cis-acting gene,
Cx, wherever they may exist in the chromosome set. Structural genes,
Sx, are assumed to play no regulatory roles, although this assumption
is not critical to the discussion. In figure 1a I have arrayed 16 sets of
triads of cis-acting, trans-acting, and structural genes separated by blanks
on the 4 chromosomes. A graphical representation of the control interactions among these hypothetical genes is shown in figure 2a, in which

CHROMOSOME I   CI T2 SI—C2 T3 S2—C3 T4 S3—C4 T5 S4—

CHROMOSOME 2   C5 T6 S5–C6 T7 S6—C7 T8 S7—C8 T9 S8—

CHROMOSOME 3   C9 TIO S9—CIO TII SIO—CII TI2 SII—CI2 TI3 SI2—

CHROMOSOME 4   CI3 TI4 SI3—CI4 TI5 TI4—CI5 TI6 SI5—CI6 TI SI6—

Figure 1b
Similar to Figure 1a, except the triads are $C(x)$, $T(x + 1)$, $S(x)$.

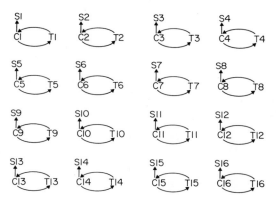

Figure 2a
Representation of regulatory interactions according to rules among genes in chromosome set of figure 1a.

an arrow is directed from each labeled gene, shown as a dot, to each gene which it affects. Thus in figure 2a C1 sends an arrow to T1 and an arrow to S1, while T1 sends an arrow to C1. A similar simple architecture occurs for each of the 16 triads of genes, creating 16 separate genetic feedback loops. By contrast, in figure 1b, each triad carries the indices (Cx, Tx + 1, Sx), while the 16th is (C16, T1, S16). This permutation yields a control architecture containing one long feedback loop, figure 2b.

The first point of figures 1 and 2 is that each spatial arrangement of cis-, trans-, and structural genes into domains along the chromosomes can be put into one-to-one correspondence with a specific wiring diagram showing which genes affect which genes. While this hypothetical example is obviously idealized, it is also true of the genomic system in contemporary cells. A spatial arrangement into domains of cis-acting, trans-acting, and structural genes corresponds to some actual wiring

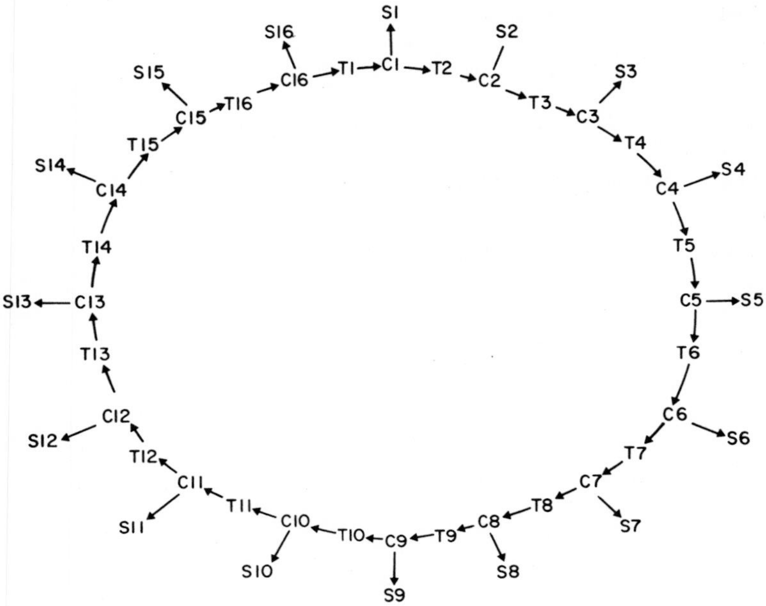

Figure 2b
Regulatory interactions among chromosome set in figure 1b.

diagram of regulatory interactions, for example, bacteriophage lambda (Thomas, 1979). Inclusion of the complete list of interacting products of each gene as components of a eukaryotic system would correspond to drawing a more complex wiring diagram.

The second obvious point of figures 1 and 2 is that alterations of the spatial arrangements of cis-, trans-, and structural genes on the chromosomes correspond to alterations in the wiring diagram.

The Scrambling Genome

It is now widely, but not universally, thought that chromosomal mutations are an important factor in evolution (Wilson et al., 1977; Bush, et al., 1977; Bush, 1980; Flavell, 1982; Dover et al., 1982). These mutations include not only tandem duplications that generate new copies of old genes but growing evidence suggests that fairly rapid dispersion of some genetic elements occurs through chromosomal mutations including transposition, translocation, inversions, and conversions. Dispersal of loci by processes such as these provide the potential to move cis- or trans-acting genes to new positions and thereby create novel regulatory connections, having implications for oncogenesis (Cairns, 1981) while also opening novel evolutionary possibilities.

Appreciation of the potential evolutionary import of chromosomal mutations is widespread (Roeder and Fink, 1980; Shapiro, 1981; Gillespie et al., 1982; Hunkapiller et al., 1982; Smith, 1982). The new issue I want to address is this: Suppose we are able to characterize with some precision the frequencies with which different DNA regions actually duplicate, or are dispersed randomly or nonrandomly in the chromosome set. Some currently unknown subset of these alters the genomic wiring diagram. Suppose for the moment we ignore any effect of selection. Then chromosomal mutations persistently scramble the genomic wiring diagram in some way consistent with what we shall eventually discover about the actual probabilities of local duplication and dispersion. In the absence of selection, other than that which may influence the chromosomal mutation rules themselves, this scrambling process will explore the ensemble of all possible wiring diagrams consistent with the constraints in the scrambling process. Thus, in the absence of further selection, we would expect the persistently scrambled genomic system eventually to display a wiring diagram whose features were typical of the ensemble. Therefore, characterizing such typical or generic properties is fundamental, for they are just the properties to be expected in the absence of selection beyond the rules of the scrambling process itself. The typical properties of the well-scrambled genome constitute the proper null hypothesis for our further analysis. I next show that the well-scrambled genome can be expected to exhibit very highly self-organized properties.

We are not yet able to characterize the ensemble of genomic systems actually explored in evolution. But a simple beginning can be made with the idealized model in figures 1 and 2. To begin study of the effects of duplication and dispersion of loci on such simple networks, I ignored questions of recombination, inversion, deletion, translocation, and point mutations completely, and modelled dispersion by using transpositions alone. I used a simple program which decided at random, for the haploid chromosome set, whether a duplication or transposition occurred at each iteration, over how long a linear range of loci, between which loci duplication occurred, and into which position transposition occurred.

Even with these simplifications, the kinetics of this system are complex. They need not be further discussed here, since the major purpose

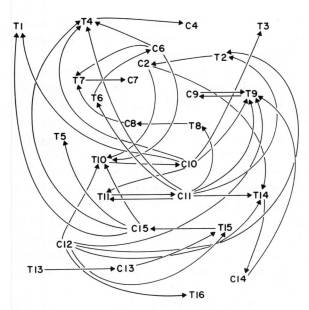

Figure 3a
Regulatory interactions from chromosome set in figure 1a after 2,000 transpositions
and duplications have occurred, in ratio 0.90:0.10, each event including 1 to 5 adjacent loci. Structural genes and fully disconnected regulatory genes are not shown.

of this simple model is to examine the regulatory architecture after
many instances of duplication and transposition have occurred. I show
the results for the two distinct initial networks in figures 3a and 3b,
for conditions in which transposition occurs much more frequently than
duplication, .90:.10, and 2,000 iterations have occurred. The effect of
transpositions is to randomize the regulatory connections in the system,
thereby exploring the ensemble available under the constraints on duplication and dispersion. Consequently, while the placement of individual genes differs, the two resulting networks (figures 3a and 3b)
look far more similar in overall architecture after adequate transpositions
have occurred than at the outset (figures 2a and 2b). This similarity,
readily apparent to the eye, reflects the fact that both networks in 3a
and 3b exhibit connectivity properties that are typical in the ensemble
explored by the model chromosomal mutations. These generic connectivity properties can be stated more precisely.

The over-simple model in figures 1–3 is still too complicated for this
initial discussion. Duplication and transposition create new copies of

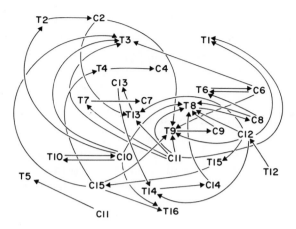

Figure 3b
Similar to figure 3a, after random transposition and duplication in chromosome set
from figure 1b.

old genes, and by their dispersal to new regulatory domains generate
both more and novel regulatory couplings among the loci. A minimal
initial approach to the study of the ensemble properties of such systems
is to study the features of networks in which N genes are coupled
completely at random by M regulatory connections. This kind of structure is termed a *random directed graph* (Berge, 1962), in which nodes
(dots) represent genes and arrows represent regulatory interactions.

To analyze such random networks I employed a computer program
that generated at random M ordered pairs chosen among N genes,
assigned an arrow running from the first to the second member of each
pair, then analyzed the following fundamental connectivity features of
the resulting wiring diagram: (1) The number of genes directly or indirectly influenced by each single gene, by following regulatory arrows
tail to head, termed the regulated *descendents* from each gene. A well-
known example of such direct and indirect genetic influence is the
sequential cascade of over 150 alterations in *Drosophila* salivary gland
chromosome gene-puffing patterns following exposure to the molting
hormone ecdysone (Ashburner, 1970). (2) The radius from each gene,
defined as the minimum number of steps for influence to propagate
from that gene to its entire battery of descendents. The mean radius
and the variance of the radius give information on how extended and
hierarchically constructed the network is. Thus long hierarchical chains
of command from a few chief genes would have high mean and variance

in the radius, while a network in which each gene influenced all other genes in one step would have low mean radius and low variance. (3) The fraction of genes lying on feedback loops among the total N genes. (4) The length of the smallest feedback loop for any gene which lies on a feedback loop.

Connectivity Properties of Regulatory Systems

The expected network connectivity features exhibit strong self-organizing properties analogous to phase transitions in physics (Erdos and Renyi, 1959, 1960), as the number of regulatory (arrow) connections, M, among N genes increases. If M is small relative to N, the scrambled genomic system would consist of many small genetic circuits, each unconnected to the remainder. As the number of regulatory connections, M, increases past the number of genes, N, large connected genetic circuits form. This crystallization of large circuits as M increases is analogous to a phase transition. An example, with $N = 20$ genes, and an increasing number of arrows, $M = 5, 10, 20, 30, 40$, is shown in figure 4. Thus, if $M = 40, N = 20$, typically direct and indirect regulatory pathways lead from each gene to most genes, many genes lie on feedback loops, etc. Results are shown in figures 5a–d for larger networks with 200 genes and regulatory connections, M, ranging from 0 to 720. As M increases past N, large descendent structures arise in which some genes directly or indirectly influence a large number of other genes. The mean descendent curve is sigmoidal (figure 5a). When $M=3N$ each gene typically is directly or indirectly connected to most other genes. The mean radius of the genetic wiring diagram is a nonmonotonic function of the ratio of M to N (figure 5b). As M increases from 0, each gene can reach its few descendents in a few steps (figure 4). As long descendent circuits crystallize in the still sparse network, when M is only slightly greater than N, the radius becomes long. As the number of connections, M, continues to increase, new short pathways connecting a gene with its descendents are formed and the radius gradually decreases (figure 4). The fraction of genes lying on feedback loops (figure 5c), like the number of descendents, is sigmoidal as M increases past N, since long descendent structures are likely to form closed genetic feedback loops as well. Finally, since the minimal lengths of feedback loops are bound above by the radius (figure 5d), these are also a nonmonotonic function of the $M:N$ ratio, reaching a maximum about where the two sigmoidal curves are steepest.

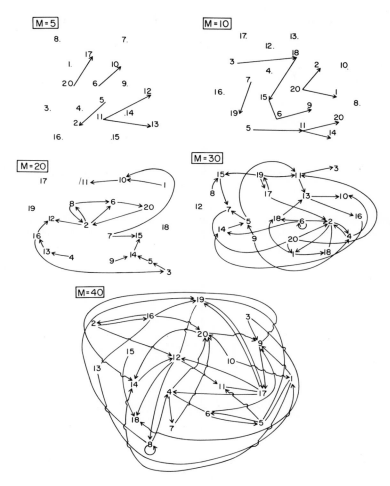

Figure 4
Random regulatory circuits among 20 genes as the number of connections, *M*, increases from 5 to 40.

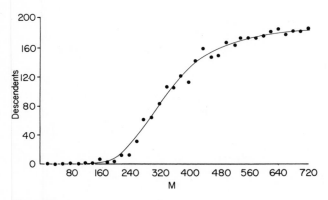

Figure 5a

Descendent distribution, showing the average number of genes each gene directly or indirectly influences as a function of the number of genes (200) and regulatory interactions, M.

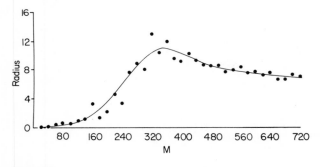

Figure5b

Radius distribution showing the mean number of steps for influence to propagate to all descendents of a gene as a function of M.

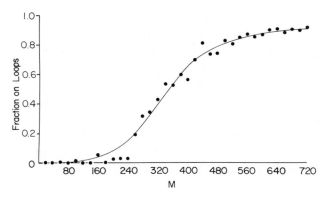

Figure 5c
Average number of genes lying on feedback loops as a function of *M*.

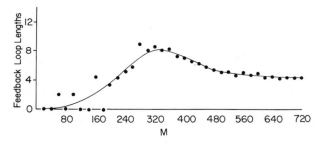

Figure 5d
Average length of shortest feedback loops genes lie on as a function of *M*.

What shall be said of this idealized model? True genetic regulatory systems presumably do not generate novel regulatory genes and novel regulatory connections in as simple a way as this primitive model. The true mechanisms shall define some more realistic and complex ensemble actually being explored in eukaryotic evolution. However, this primitive model already points out a number of fundamental facts. First, the statistical properties of such ensembles are highly robust, just as the statistical properties of fair coin flipping ensembles are robust. Just as the expected fraction of heads is 50 percent with fair coins, so the expected connectivity properties of scrambled genetic regulatory systems are robust, and given the ensemble, characterizable. Second, the generic connectivity properties of the ensemble are not featureless, but highly structured. In the current primitive example, given 10,000 genes, and 30,000 regulatory connections among them, very strong statements can be made about the expected connectivity properties of the regulatory network. To restate the now obvious point, if we can specify how chromosomal mutations actually generate new genes and novel connections, then we can characterize the ensemble actually being explored by evolution. The highly structured generic properties of the ensemble are a proper null hypothesis characterizing in detail the statistical features we would expect to find in eukaryotic genomic systems in the absence of selection. If we do not find those features, then either we have generated the wrong ensemble because we do not understand how the genome is actually scrambling, or other forces are causing the genomic wiring diagrams to deviate from the generic. Chief among potential forces is selection.

Generic Properties, Mutation, Selection, and Limits of Adaptation

Underlying the approach I propose are the interrelated concepts of the statistically expected properties of the ensemble being explored in evolution, the effects of mutation and selection, and the limits of adaptation attainable in a population. These can now be preliminarily summarized. I begin by informally borrowing Maxwell's Demon from physics. Consider the well-known demon and two isolated and initially isothermal gas-containing boxes joined by a window. The demon allows faster molecules into the right box. As that box becomes warmer, increased pressure within it counters the demon's efforts. If the demon is of finite and modest efficiency, in due course his efficiency equals the back pressure and the system comes to a steady state displaced somewhat from thermodynamic equilibrium. The deviation from equilibrium reflects the limited capacity of the demon to direct the faster molecules into the right box. Further, if the steady state is close to thermodynamic equilibrium, many statistical properties of the selected system will be explicable from the statistical features of the unselected equilibrium ensemble properties.

The demon analogy carries directly into the evolutionary problem. The generic properties of the genetic ensemble, with respect to the wiring diagram described, are fundamental, for those properties are exactly the (equilibrium) features to be expected in the absence of selection. The generic properties are simultaneously available in the ensemble to be exploited by selection; yet also, since mutations drive partly selected systems back toward generic properties, those generic properties are a constraint to be overcome by selection. In turn, mutations drive systems toward generic properties but also supply the population variance upon which selection can act. Selection is analogous to Maxwell's demon, able to enrich the representation of the more desired genomic systems in the population, until the back pressure toward generic properties due to mutation balances the effect of selection. That balance is the limit achievable by selection. If the balance lies close to the unselected mean, then the underlying generic properties of the ensemble will in large measure account for the properties we see. Then within the ensemble actually explored in evolution, its highly structured generic statistical properties will function as normal forms or ahistorical universals, widely shared among organisms, in spite of selection.

Selection for a Unique Arbitrary Wiring Diagram

A central issue for this ensemble pattern of inference is whether selection is sufficiently powerful to achieve and maintain any arbitrary unique wiring diagram in the ensemble throughout the population. Might its properties deviate extensively from the generic properties of the ensemble? In this section I show for fixed selection rules that as the complexity of the wiring diagram to be maintained in the population by selection increases, a limit is reached beyond which a unique, precisely chosen system cannot be maintained. Further, the maximally selected network falls ever closer to the unselected mean properties of the ensemble. Hence, for sufficiently complex genetic systems, the ensemble pattern of inference should be useful. Beyond patterns of inference, however, this is a fundamental biological question about the limits of the capacity of selection to attain high precision and complexity in genetic regulatory systems in the face of mutational disruption of the selected system.

   To start investigating this question, I have begun with the simplest class of population selection models, based on the Fisher-Wright paradigm (Ewens, 1979). Imagine a population of initially identical haploid organisms with N regulatory genes connected by a total of T regulatory arrows. Suppose that some arbitrary specific wiring diagram assigning the T arrows among the N genes is "perfect," and that at each generation mutations alter the wiring diagram by probabilistically reassigning the head or tail of arrows to different genes. Assign a relative fitness to each organism which depends upon its deviation from the perfect wiring diagram, such that the probability of leaving an identical offspring in the next discrete generation is proportional to relative fitness.

Perhaps the simplest form of this general model is

$$W_x = (1 - b)\left(\frac{G_x}{T}\right)^\alpha + b, \tag{1}$$

where W_x is the relative fitness of the xth organism in the population, G_x is the number of good connections among its components, T is the total number of regulatory connections, b is a basal fitness, and the exponent α corresponds to the three broad ways fitness might correlate with the fraction of good connections. For $\alpha = 1$, fitness is linearly proportional to the number of good connections. For $\alpha > 1$, fitness drops off sharply as the number of good connections, G, falls below T. This corresponds intuitively to the case where virtually all connections must be correct for proper functioning, and any defect yields sharp loss of function. For $0 < \alpha < 1$, small deviations from perfect ($G = T$) would yield only slight loss of fitness, which would fall off increasingly rapidly for larger deviations of G below T. This corresponds to a broad region of high fitness near perfect with wide toleration of deviations in wiring diagrams. These three classes of fitness seem fundamental; more complex patterns would be built up from these three.

I have carried out an initial computer investigation of this model for small T and small population size. At each generation mutations (at rate μ per regulatory connection per generation) randomly altered wiring diagrams in a population of 100 haploid organisms, the fitness of each was calculated, organisms were sampled with replacement and left progeny in the next generation with a probability equal to their fitnesses. Sampling continued until 100 offspring were chosen, thus constituting the next generation; then the process was iterated. The initial population of organisms was identical, and either perfect or constructed randomly, hence generic. I sought, in particular, the steady states of this system, reflecting stationary population distributions of the number of correct connections per organism due to the balance of mutation and selection forces.

A general implication of this simplest of models is that for sufficiently simple wiring diagrams, selection is powerful enough to maintain the population near any arbitrary perfect wiring diagram. But with increasing complexity of the wiring diagrams, the steady state balance struck between selection and mutational back pressure shifts toward generic.

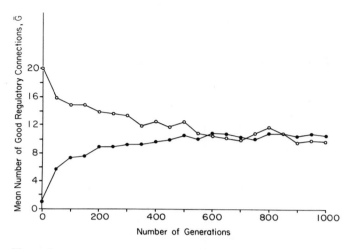

Figure 6a

Population selection for 1,000 generations after initiation at $\bar{G} = T = 20$, and $\bar{G} = 1$. $\bar{G}$, mean number of good connections per organism. Mutation rate per arrow end $\mu^* = 0.005$. $\alpha = 1$, $b = 0$, $P = 0.95$.

In figure 6a I show sumulation results for a population of 100 organisms, each with 20 genes connected by 20 arrows. At over 1,000 generations a population that initiated at the prefigured perfect wiring diagram gradually decreases the mean number of good connections $\bar{G}$, from 20 to about 10. By contrast, a population of initially identical organisms constructed at random with respect to the perfect wiring diagram, hence with an average of about one good connection, gradually increased the mean number of good connections above this generic value, to about 10. Thus under these conditions populations exhibit a single globally stable stationary distribution with $\bar{G} = 10$, reflecting the balance between mutation and selection.

As the number of connections, T, to be specified increases, the stable stationary distribution shifts smoothly from perfect closer to generic. Figure 6b shows results from simulations in populations as T ranges through values from 2 to 24, but all other parameters remain fixed. The effect of increasing the mutation rate, μ, was also investigated. With all other parameters fixed and at low mutation rates, selection can maintain any arbitrary perfect wiring diagram in all members of the population, and, while increasing the mutation rate smoothly, shifts the globally stable stationary state closer to generic (figure 6c).

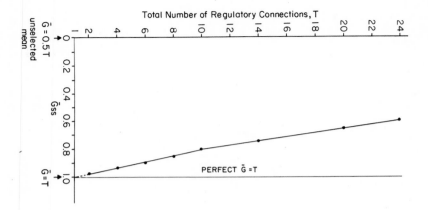

Figure 6b
Proportional shift of selected stationary state, $\bar{G}_{SS}$, from $\bar{G} = T$, labeled 1.0, toward unselected mean $\bar{G} = 0.5T$, labeled O, as T increases. $\mu^* = 0.005$, $P = 0.5$, $\alpha = 1$, $b = 0$.

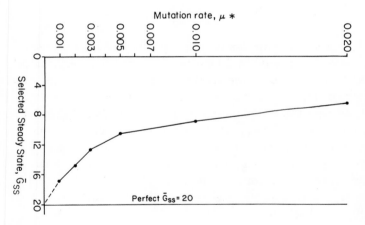

Figure 6c
Shift of selected steady state, $\bar{G}_{SS}$, toward unselected mean, $\bar{G} = 1$ as μ^* increases. $T = 20$, $P = 0.95$, $\alpha = 1$, $b = 0$.

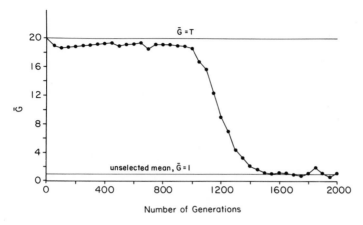

Figure 7
Selection system exhibiting two stationary states near perfect, $\bar{G} = T$, and unselected, $\bar{G} = 1$, with fluctuation driven transition. $T = 20$, $P = 0.095$, $\alpha = 10$, $b = 0.5$, $\mu = 2\mu^* = 0.01$.

Multiple Stationary States Between Generic and Perfect

The fitness rule in this simple model has a single perfect wiring diagram; hence the fitness surface has a single maximum in the space of possible wiring diagrams. It is therefore interesting that through mutation and selection the population system can have either one or two stationary states between generic and perfect. Simulation results suggest that populations have a single globally stable stationary distribution under two general parameter conditions, $\alpha = 1$ or $b = 0$. Here increasing the complexity of the network to be maintained causes a smooth shift of the selected stationary distribution toward the unselected generic properties of the ensemble. In contrast, for an appropriate choice of parameters outside these conditions the selection system can display two stationary distributions: one close to perfect, one close to generic, with fluctuation-driven transitions between them. Figure 7 shows the temporal behavior of a population (with $\alpha = 10$ and $b = 0.50$) initiated in the perfect state, which remains in that steady state for 1,000 generations and then undergoes a rapid transition to the steady state near the generic state, $\bar{G} = 1$, where it thereafter remains.

It is particularly interesting that this model here exhibits what might be called a complexity catastrophe. As the complexity of the network to be maintained, that is, T, increases, selection can maintain the pop-

Maximally Adapted Wiring Diagram Achievable by Selection

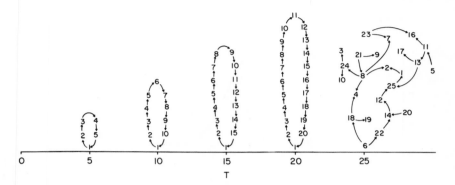

Figure 8

Maximally adapted wiring diagram achievable in a population by selection ($\alpha = 10.0$, $b = 0.5$, $\mu = 0.01$) for a single closed regulatory loop. T is the number of regulatory connections. By $T = 25$ the single loop cannot be maintained and the maximally adapted wiring diagram achievable falls toward the ensemble generic properties. In this simulation the number of genes, N, $= T$. Since $N = T$, and $P = (N-1)/N$, P changes in these simulations, but the destabilization of the nearly perfect state also occurs as T increases while P remains constant.

ulation in the nearly perfect stationary state until it decreases in stability and entirely disappears, causing the population to jump suddenly to the remaining single globally stable steady state close to generic. Thus a small increase in the complexity to be maintained, T, can cause a catastrophic change in the properties of the maximally adapted system attainable (figure 8).

Analytic Approaches

The question to be assessed is whether the balance struck between mutation and selection will result in organisms whose genetic wiring diagrams are close to the typical properties of the well-scrambled genomic system or arbitrarily different from those properties. Although simulation results on small systems are useful, analytic insight allowing consideration of large networks is necessary. A full analytic treatment of this model is not yet available, but I report in more detail elsewhere (Kauffman, 1983) an approach based on the mean number of good connections per organism in the population, $\bar{G}$, and the variance in the number of good connections per organism in the population, $\sigma^2 G$.

For the full fitness law, $W_x = (1 - b)(G_x)^\alpha + b$, where $1 \geq b \geq 0$ reflects a basal fitness and P is the probability that any arbitrary connection is bad. The differential equation describing rate of increase in the mean number of good connections per organism, $\bar{G}$ is

$$\frac{d\bar{G}}{dT} = \frac{\alpha(1 - b)G^{\alpha-1}\sigma^2 G}{(1 - b)^\alpha + bT^\alpha} - \mu(TP + \bar{G} - T). \tag{2}$$

The first term on the right-hand side of the equation is the effect of selection, Se, attempting to increase G, and the second term is the restoring force, Rf, of mutation, at rate μ, decreasing $\bar{G}$ toward the value of $\bar{G}$ which would occur without selection, $\bar{G}_\gamma = T(1 - P)$.

This equation accounts for the simulation behavior. Note that the mutational restoring force, Rf, not only decreases the number of good connections, $\bar{G}$ but as in the demon analogy, this back pressure increases monotonically as $\bar{G}$ increases toward T. The effect of selection increasing $\bar{G}$ depends on parameters α, b, and $\sigma^2 G$. For $\alpha = 1$, that is, where fitness is linearly proportional to the fraction of good connections, or for $b = 0$ (no basal fitness) the selective term, Se, decreases monotonically as $\bar{G}$ increases. For parameter values where the selective force is greater than the restoring force for all values of $\bar{G}$ less than T, the curves representing the strength of Rf and Se do not intersect in the interval $0 \leq \bar{G} \leq T$, and selection is sufficiently powerful to maintain any unique arbitrary perfect wiring diagram in almost all members of the population. For parameter values where the Rf and Se curves intersect in the interval $0 \leq \bar{G} \leq T$ (figure 9a), they must do so at a single point, representing a single globally stable stationary population distribution of fitness around $\bar{G}$. This reflects the limited fitness achievable by selection due to the balancing back pressure of mutation.

As noted before, even with a single maximum of the fitness surface at the perfect wiring diagram, this system can exhibit two stable stationary distributions: one near the unselected generic level of G, the other near perfect $\bar{G} = T$. As shown in figure 9b, for values of α sufficiently large, such as 10, and b greater than 0, such as 0.5, the selective curve, Se, does not decrease as $\bar{G}$ increases. Rather it increases at more than a linear rate, hence can cross the mutational restoring force curve Rf at two points. The point closest to generic, $G1$, is a stable stationary state. The point closer to perfect, $G2$, is unstable and repels the population in either direction. The perfect state, $\bar{G} = T$, is a reflecting boundary. Therefore, if the population is above the unstable steady state, $G2$, it becomes trapped at a nearly perfect state between $G2$ and

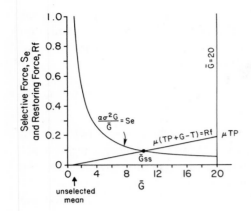

Figure 9a
Restoring force, *Rf*, tending to decrease $\bar{G}$ toward unselected mean and selective force, *Se*, to increase $\bar{G}$. $\bar{G}_{ss}$ is the selected steady state. $T = 20$, $P = 0.95$, $\alpha = 1$, $b = 0$, $\mu = 2\mu^* = 0.01$. μTP is *Rf* at $\bar{G}=T$.

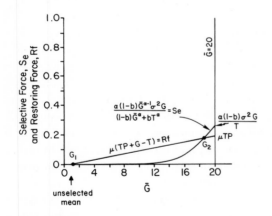

Figure 9b
Selective force, *Se*, increasing with $\bar{G}$ for $\alpha = 10$, $b = 0.5$, $T = 20$, $P = 0.95$, $\mu = 2\mu^* = 0.01$. G1 is the stable steady state, and G2, the unstable steady state. μTP is *Rf* at $\bar{G} = T$. $[\alpha(1 - b)\sigma^2 G]/T$ is *Se* at $\bar{G} = T$.

perfect. Thus the population shows two stable stationary states, with fluctuation-driven transitions between them (figure 7).

One major feature of the analysis at this point is that if the network to be achieved is sufficiently simple, selection can maintain essentially the entire population at any arbitrary perfect network. Simulation results show, however, that as the total complexity to be achieved, T, increases, or the mutation rate, μ, increases, selection is unable to maintain the population near any unique perfect network. Analysis of the effects of increasing the total complexity, T, or the mutation rate, μ, on the system characterized by Equation 2, requires assessment of the change they cause in $\sigma^2 G$. Simulation results suggest that as T increases, $\sigma^2 G$ increases proportionally, but that as μ increases, $\sigma^2 G$ increases only proportionally to the square root of μ, $\mu^{1/2}$.

The capacity of the system to maintain a population at a perfect wiring diagram requires that the selective force in that state be greater than the restoring force. Substituting $\bar{G} = T$ into Equation 2 and using the approximation $\sigma^2 G = KT\mu^{1/2}$ yields

$$\frac{d\bar{G}}{dt} = \frac{\alpha(1 - b)KT\mu^{1/2}}{T} - \mu TP \geq 0. \tag{3}$$

Persistence of the perfect state requires $Se \geq Rf$, or simplifying,

$$\frac{\alpha(1 - b)K}{P} \geq \mu^{1/2}T. \tag{4}$$

As T increases, the selective force is roughly independent of T, but the restoring force due to mutation increases proportionally to T. Therefore as T increases, selection can at first maintain the population in a nearly perfect stationary state. As the complexity of T increases further, the selection force at the perfect state, $\bar{G} = T$, falls below the mutational restoring force, Rf, and the system shifts to a steady state below $\bar{G} = T$. Under parameter conditions where Se decreases monotonically as $\bar{G}$ increases, that is, $\alpha = 1$ or $b = 0$, an increase in T causes a smooth shift in the single best achievable globally stable steady state $\bar{G}_{ss}$ toward the unselected generic $\bar{G}_\gamma = T(1 - P)$. Under conditions with two steady states (figure 9b), increasing T leads to a bifurcation in which the unstable steady state $G2$ disappears, and the system falls catastrophically to the near generic stationary state $G1$. Hence the system exhibits a complexity catastrophe as T increases. The basic result of this investigation, therefore, is that, with other parameters of the selection system fixed, and as the complexity of the system to be maintained increases, the system inevitably approaches the statistically typical properties of the ensemble from which it is drawn.

Compensation for Increasing Complexity by Decreasing Mutation Rate

The maximum complexity wiring diagram, T_{max}, which can be maintained with precision, depends not only on T but also on the remaining parameters. Therefore an increase in T can be balanced by compensatory decrease in the mutation rate, μ, or increase in the selective exponent α. I defer discussion of increasing α. With respect to mutation, Equation 4 suggests that the perfect state can be maintained if the term $\mu^{1/2}T \leq C$, where $C = \alpha(1 - b)K/P$, or if

$$T \leq \frac{C}{\mu^{1/2}}, \quad \text{for } \mu \leq \frac{K^2}{T^2}. \tag{5}$$

Thus the perfect state can be maintained as T increases if μ decreases inversely proportionally to T^2. An increase in the complexity of T by a hundred fold would require a decrease in mutation rate by a factor of 10^{-4}. Two immediate features of this result are obvious. First, the requirement that mutation should decrease as the square of the number of connections to be maintained makes such compensation increasingly difficult as T increases. Second, a minimum mutation rate is attainable. With other parameters fixed, such as α, this places a maximum bound on the complexity of T, (T_{max}) which can be maintained with precision by selection.

The analysis given here is an attempt using standard population genetics techniques. Therefore a particularly fascinating and important feature of the inverse relation between T^2 and μ is that it appears to be fairly general, and arises in classical population genetic models, with direct bearing on the example of genetic wiring diagrams. Classical single-locus theory considers two alleles, A1 and A2, in diploids with genotypes A1A1, A1A2, A2A2, mutation rate v from the more favored A2 to A1, and mutation at rate u from the less favored A1 to A2. If both alleles are equally fit, at equilibrium the frequency, p_x, of the less favorable allele A1 in the population is $v/(u + v)$. In the presence of selection against A1, where A1A1 has fitness $1 - S$ [or more formally $1 + (-S)$], A1A2 has fitness $1 - S/2$, and A2A2 has fitness 1, the frequency of A1 in the population is given by the confluent hyper-geometric series expression:

$$p_x = \frac{u}{u + v} \cdot \frac{M(4Nv + 1, 4Nu + 4Nv + 1, 2NS)}{M(4Nv, 4Nu + 4Nv, 2NS)}, \tag{6}$$

where N is the number of individuals in the population, and $M(\cdot)$ refers to terms in the confluent hypergeometric series. If the term $2NS$ is large and negative, requiring S negative and not too close to 0, then Equation 6 simplifies to

$$p_x = \frac{2v}{|S|}, \tag{7}$$

where $|S|$ is the absolute value of S. The frequency of the less favorable allele per individual becomes

$$f_x = \frac{2N \cdot 2v}{N|S|} = \frac{4v}{|S|}. \tag{8}$$

Single-locus theory can be directly extended to consider T independent loci, each with two alleles and each contributing additively and independently to the total fitness of the system. To ensure that the normalized maximum fitness remains 1 as T increases, this requires that the fitness contribution per locus, S' decrease, $S' = S/T$. Assuming the forward and back mutation rates at all loci are v and u, substitution yields for the fraction of less favorable alleles per individual

$$f_x = \frac{4vT}{|S|}. \tag{9}$$

Consequently the actual expected number of less favorable alleles, C, is

$$C = \frac{4vT^2}{|S|}. \tag{10}$$

This implies

$$v = \frac{C|S|}{4T^2}. \tag{11}$$

Equations 10 and 11 imply that to hold the total number of unfavorable alleles at some constant number, C, per individual as the total number of loci, T, increases, requires that the mutation rate v decrease inversely to T^2. The similarity to equations 4 and 5 is obvious. In fact it is straightforward to interpret good and bad alleles at T loci in terms of regulatory connections in a wiring diagram where, however, the two loci at each end of each connection must be good. This raises the expected number of bad connections to $8vT^2/S$, which is minimized by $|S| = 1$ to

$$C = 8vT^2. \tag{12}$$

This interpretation corresponds most closely to the haploid model where fitness is linearly proportional to the number of good connections, $\alpha = 1$ and where there is no basal fitness, $b = 0$, $W_x = (G_x/T)$. Thus fitness ranges linearly from 0 to 1, with each good connection adding $S' = 1/T$ to the total fitness.

A basic implication, then—following securely from the classical genetic example corresponding to the simplest fitness law, $\alpha = 1$, $b = 0$, and with substantial plausibility even for $\alpha > 1$, $b > 0$—is that for a fixed mutation rate the number of bad connections in a genetic wiring diagram will initially increase as T^2. As T continues to increase, the system eventually approaches the unselected, well-scrambled equilibrium system where the fraction of bad connections is constant; hence their numbers increase linearly with T.

These results allow an estimation of the number of mistaken regulatory connections to be expected in plausibly sized genetic regulatory systems. A minimum estimate of the number of transcribed sequences in a higher eukaryote is on the order of 20,000 to 100,000. The frequency of chromosomal or other mutations that alter regulatory couplings is unknown, but 10^{-6} is plausible. Consider a modest-sized regulatory system with 1,000 regulatory genes, each connected to 10 other genes, thus comprising 10,000 regulatory connections. Equation 12 suggests that such a network would have on the order of 800 mistaken connections. Were each of the 1,000 genes connected to 30 genes, about 7,200 connections would be mistaken; that is, almost 25 percent of the connections would be incorrect. I show briefly later that 800 or 7,200 misconnections would suffice to ensure the occurrence of generic connectivity properties in such a genomic regulatory system.

Ensemble Generic Properties under Four Conditions

Do we, then, actually expect the genomic wiring diagram in contemporary cells to resemble the ensemble genetic properties of the well-scrambled genome? From the foregoing discussion it is clear that if the wiring diagram is sufficiently simple selection can maintain arbitrary and precise wiring diagrams in contemporary organisms. There are four major cases where one would expect to observe an approach to the generic properties.

1. The generic features are themselves selected for. There is no obvious reason why such features should not be selected for, so I ignore this case.

2. The generic statistical features of such wiring diagrams are selectively neutral. This may be unlikely. Were it true, then those features would of course occur. I do not discuss this case further.

3. The generic features are sometimes selected against, but the total complexity of the system precludes maintenance of precision. As discussed earlier, for fixed fitness laws and a minimized mutation rate as T increases, this collapse toward generic must eventually occur. Why might evolution lead to wiring diagrams so complex that their precision cannot be assured? One important possibility is that although both the fraction and the total number of bad connections increases, and the fraction of good connections decreases, the total number of good connections increases. It may be advantageous to sacrifice precision to gain this increase. Indeed, the implication of the preceding discussion is that a contemporary eukaryote with 50,000 genes whose expression requires regulation is very probably far too large for the additive selection to maintain the regulatory system with precision.

4. Maintenance of adequate population variance in a heterogeneous environment may require a substantial fraction of the population to have wiring diagrams that approach generic features. This fourth condition may be of major importance and requires further discussion.

In my earlier analysis I delayed consideration of the selective exponent α. Consider the fitness law $W_x = (1 - b)(G_x/T)^\alpha + b$. For very large α, $b = 0$, this law asserts that addition of any new good connections leads to an organism very much more fit than all others. Thus its progeny will almost instantaneously take over the population. For arbitrarily large α, the population will inexorably step upward to arbitrarily complex, perfect wiring diagrams. Large α is like a powerful Maxwellian demon. At least two factors may bound α from becoming arbitrarily large. First, it may not often be the case that adding the next good connection actually increases fitness so dramatically. More important, however, is the high price to be paid for such stringent selection in loss of population variance.

The simplest model considered here had a single perfect wiring diagram, hence a single global maximum in the fitness surface over all possible wiring diagrams. Suppose instead, with Sewell Wright, that many fitness peaks occur and that the mean distance between peaks, in terms of numbers of mutational events, that is, bad connections, needed to cross from one peak to another is C^*. If the selective exponent, α, is very large, the population will be held in a narrow distribution around a local peak. If the environment is spatially or temporally patchy, and transition to a new peak is needed, the population cannot do so unless it has sufficient variance to have members across the valley between the neighboring peaks. If the mean distance between peaks is C^*, on average the distance from one peak to the valley is $C^*/2$. Too narrow a population distribution around the peak precludes the transition.

These are of course old ideas, underlying the classical concept that maintenance of population variance is advantageous. The new issue I want to raise is this: if populations maintain enough variance that some members are $C^*/2$ from arbitrarily improbable fitness peaks, will those less fit variants exhibit ensemble generic properties? A concrete example suggests that the answer will often be yes.

Consider a genetic wiring diagram with 1,000 regulatory genes, separated into 10 independent subwiring diagram blocks, each containing 100 genes. Within each independent block let each gene be connected to 10 others. Thus there are 10,000 total connections. This wiring diagram among 1,000 genes is obviously very improbable in the ensemble of all possible wiring diagrams with 1,000 genes and 10,000 regulatory connections. In particular, its connectivity properties are highly atypical. Each gene is connected directly or indirectly only to 100 among the 1,000. Were the network randomly connected, each would directly or indirectly connect to virtually all other genes.

Since there are 10 such sets, each with 100 genes, about 20 random mistaken connections linking single genes in one block with single genes in another block would suffice to interconnect all 10 blocks of 100 genes. Then genes in each block would directly or indirectly connect to virtually all 1,000 genes, rather than to a mere 100. Thus a small number of mistakes largely restores a highly atypical connectivity feature toward generic. As noted before, with a mutation rate $\mu = 10^{-6}$, a system with 10,000 connections would be expected to have about 800 mistaken connections, far exceeding the 20 mistakes needed to restore this generic connectivity property. To continue this hypothetical example, if each regulatory gene connects to 30 others rather than 10 others within its own independent set of 100 genes, thereby increasing the number of regulatory connections to 30,000, 20 mistaken connections would still interconnect the 10 blocks of 100 genes and restore the network dramatically closer to generic connective features with respect to the number of descendents of each gene. Equation 12 suggests that in this case 7,200 connections would be incorrect.

The example is hypothetical but captures the fact that if the selected system is enormously atypical of the underlying ensemble in some respect, then rather few mistakes often induce a very substantial restoration of ensemble typical properties. It is not hard to construct examples where the numbers of mistakes required are constant, or only increase slowly, as the complexity of the selected system increases. Therefore, if the distance between adaptive peaks, C^*, is substantially larger than that required to restore particular generic features, then meeting the necessity to remain adaptive and ensure adequate population variance for transitions to neighboring peaks will also ensure that populations in the adaptive valleys exhibit those generic features.

Selective Adaptation of Dynamical Behavior of Genetic Regulatory Systems and the Limits of Selection

This discussion has focused entirely on the wiring diagram architecture of genetic regulatory systems under the drives of mutation and selection. Although difficult to visualize, these architectural features are easier to describe than the dynamical behavior of genetic regulatory systems, reflected in expression of coordinated sets of genes, which underlie ontogeny. Such analysis is truly the most important goal of the ensemble approach I advocate here. Just as the wiring diagrams of regulatory networks subject to certain classes of mutants explore a definable ensemble of accessible regulatory wiring diagrams, so the classes of mutants, which alter the regulatory rules governing the activity of any gene as a function of those molecular variables that are its inputs, also explore specific and definable ensembles of dynamical regulatory systems. Because distinct ensembles have distinct generic properties, which I will discuss shortly, the task of characterizing the ensemble that contemporary organisms are in fact exploring is a critical one. As suggested by the Maxwell Demon analogy and the preceding analysis, selection will be expected to be unable to avoid those generic properties if the regulatory system is sufficiently large; thus the generic properties of the ensemble we actually are exploring may constitute genuine biological universals.

The genome is a higher eukaryote has on the order of 100,000 genes. Directly or indirectly these regulate one another's expression. To study such complex systems, one useful approach is to idealize the activity of any gene as 1 or 0, hence to model the genetic regulatory system as a complex binary element, or switching automaton. If a model genome has N genes, it then has 2^N possible combinations of gene activities, each constituting one state of the system. The regulatory network transforms each state into a successor state. Over a sequence of time moments the network passes through a trajectory of states or patterns of gene expression. The number of states is finite; hence the system must return to a state previously encountered. For the simplest automata models, all genes switch values simultaneously; thus the system is deterministic. Once the system reenters a pattern of gene expression previously encountered, it cycles recurrently among this reentrant state cycle. If released from a different initial state, the system may flow to this first-state cycle or into a distinct state cycle. Thus state cycles are the asymptotic attractors of such networks. After transients die away, the system will remain cycling in one or another of its state cycle attractors. The repertoire of attractors constitutes the repertoire of alternative recurrent patterns of gene expression open to such model genomic systems.

It is natural to model each cell type of an organism as one such recurrent pattern of gene expression or a dynamical attractor of the genomic system. If so, then the lengths of state cycles (the numbers of states on the recurrent sequence) predict how tightly clustered the patterns of gene expression corresponding to one cell type are expected to be. The number of state cycle attractors predicts the number of distinct cell types the genomic system can express. The stability of attractors to fluctuations gives indications about the expected stability of cell types to environmental fluctuations and noise while behaving. If an attractor is a cell type, differentiation is passing from one attractor to another. The number of neighboring attractors reachable from each attractor by suitable signals carries implications about how many neighbor cell types each cell type can differentiate into directly.

These dynamical properties differ in different ensembles of model genetic regulatory systems. Briefly, three alternative ensembles have been studied. (1) Model genomes in which each gene is directly regulated by a large number of other genes, K, and the Boolean switching rule governing its behavior can be chosen ad libitum from among the 2^{2^K}

possible Boolean rules of K input variables. (2) Each gene is regulated by a large number, K, of other genes, but the governing Boolean rules are limited to threshold functions. Thus each gene has excitatory and inhibitory inputs from other genes and is active at the next moment if the weighted sum of inhibition exceeds some threshold. (3) Genes have rather few inputs, K, from other genes or are restricted, if K is large, to a canalizing subset of Boolean functions. Canalizing functions capture in the Boolean idealization the actual rules that appear to govern most regulated prokaryotic and phage genes (Kauffman, 1969, 1974, 1983).

Of these three ensembles the last generically exhibits the most highly ordered dynamical behavior. Further, without any selection, these systems generally exhibit behaviors paralleling many features of contemporary organisms. In particular, the lengths of state cycles are a mere N, and the numbers of distinct state cycles are also near N. Thus a regulatory system with 10,000 genes and $2^{10,000}$ patterns of gene expression would settle into about 100 alternative recurrent patterns, or cell types, each processing through about 100 similar states of coordinated gene expression. This suggests that the numbers of cell types in organisms should rise as about a square root function of the number of genes, that is, less than linearly with genomic complexity. This appears to be approximately correct (Kauffman, 1969, 1974). Expected differences in patterns of gene expression in different cell types of one model organism are close to those actually observed, so is the predicted existence of a large fraction of the genes ubiquitously expressed in all cell types (Hough et al., 1975; Bantle and Hahn, 1976; Hastie and Bishop, 1976; Kleene and Humphries, 1977; Chikaraishi et al., 1978). Under perturbation, which reverses the activity of any single gene at a time, such model cell types are stable; that is, cell types are homeostatic. Critically each cell type can differentiate directly to only a few alternative cell types, with appropriate signals. Therefore for any cell type, such as the zygote, to differentiate into many cell types, it must do so along branching pathways of differentiation. Mutations deleting or adding genes to such systems cause only modest changes in patterns, 5 to 15 percent, of gene expression in the remaining system, paralleling alteration of protein expression patterns seen in single mutants.

Analysis of the limits of selection for this and the other ensembles has just begun. Since the number of mistaken connections, with a linear fitness rule, rises as the square of the number of connections in the wiring diagram, so too will the number of mistaken control rules at loci rise as a square of the number of regulated loci, for a fixed mutation rate. The analysis of selection and of its limits shows that for fixed selection rules a sufficiently small genomic regulatory system can be maintained by selection of any required uniquely highly adapted regulatory behavior. But as the size of the selected system increases, the number of mistakes will accumulate. For sufficiently complex systems selection will be unable to avoid the generic properties of the ensemble of regulatory systems explored by mutation. Thus in the introduction I remarked that in all metazoans and metaphytons each cell type differentiates directly into only a few neighboring cell types, thereby requiring branching pathways of differentiation in all ontogenies. If evolution is exploring the ensemble of regulatory systems characterized by the low K canalizing ensemble, then this universal feature of metazoans and metaphytons almost certainly does not reflect selection but an unavoidable generic property of the ensemble of regulatory systems from which we are all drawn. In general, the many strong parallels between generic dynamical features of the canalizing ensemble and contemporary organisms encourages the suspicion that many of those features are unavoidable by selection, hence ubiquitous.

Recapitulation

The main outlines of this discussion can now be recapitulated and placed in a slightly broader perspective. Evolutionary theory has grown almost without insight into the powerful self-organizing properties of complex systems, whose features are just beginning to be understood. I have carried out this discussion largely in terms of the highly ordered connectivity properties of well-scrambled genetic regulatory networks. An initial motivation for these inquiries lay in the discovery that such scrambled networks also exhibit extraordinarily coordinated patterns of gene expression paralleling many features of contemporary cell differentiation (Fogelman Soulie et al., 1982; Kauffman, 1969, 1974, 1982), including the implication that each cell type differentiates directly into only a few neighboring cell types. Other powerful self-organizing properties appear to arise with respect to rhythmic phenomena, and the onset of spatial structuring in reaction-diffusion, or viscoelastic models of morphogenesis (Murray, 1981).

The fundamental problems are not only to understand such self-organizing properties but to clarify their interrelation to selective forces, design principles, and historical accidents in evolution. Inclusion in this task of systems with strongly self-organizing properties raises new conceptual problems and opportunities. Those properties are themselves generators of order, relieving selection of that mantle which it has borne. On the other hand, selection either makes use of those self-organizing properties or opposes them. We have as yet little understanding of what it might mean for such internal self-organizing features of organisms to interact with selective forces, how to characterize such interactions theoretically, and how to assess them experimentally.

The particular example considered here focused on the concept of the evolutionary ensemble of genetic regulatory systems, that is, the ensemble of systems accessible in principle under the actual forms of mutation that alter the genomic regulatory system. Such ensembles naturally have statistically robust features with respect to connectivity properties and dynamical properties underlying coordination of gene expression. Those generic properties are therefore available to selection and, if opposed by selection, constitute an internal constraint that selection must continuously overcome. Such generic properties will actually be seen in contemporary organisms if (1) selection occurs for those properties; (2) selection is neutral for those properties; (3) selection opposes those properties, but the regulatory system is sufficiently complex that the balance struck between selection and the generic properties is reasonably close to those generic properties; (4) population variance in heterogeneous environments requires that at least the less fit variants will exhibit generic features.

A strength of an ensemble theory is that it allows major features of complex systems to be predicted and explained without requiring the system to be analyzed fully. The weakness of any such theory is that it predicts expected distributions, not particulars. Ultimately the route to testing such ensemble theories must confirm the expected distributions. If in fact such ensemble-generic features are widespread, they are candidates for genuine biological universals.

Note

It is a pleasure to acknowledge fruitful conversations with Drs. Vahe Bedian, Jo Darken, Warren Ewens, Thomas Schopf, Montgomery Slatkin, Alan Templeton, David Wake, and William Wimsatt. This work was partially supported by ACS CD-30, ACS CD-249, NSF PCM 8210601, and NIH GM 22341.

References

Abraham, I., and W. W. Doane, 1978. Genetic regulation of tissue-specific expression of *Amylase* structural genes in *Drosophila melanogaster*. *Proc. Natl. Acad. Sci. USA* 75:4446–4450.

Alberch, P., 1982. Developmental constraints in evolutionary processes. In *Evolution and Development*, J. T. Bonner, ed. Berlin, Heidelberg, New York: Springer-Verlag, pp. 313–332.

Ashburner, M., 1970. Puffing patterns in *Drosophila melanogaster* and related species. In *Developmental Studies on Giant Chromosomes: Results and Problems in Cell Differentiation*, vol. 4, W. Beerman, ed. Berlin, Heidelberg, New York: Springer-Verlag, pp. 101–151.

Bantle, J. A., and W. E. Hahn, 1976. Complexity and characterization of poly-adenlylated RNA in the mouse brain. *Cell* 8:139–150.

Berge, C., 1962. *The Theory of Graphs and Its Applications*. London: Methusena.

Bishop, J. O., 1974. The gene numbers game. *Cell* 2:81–86.

Bonner, J. T., ed., 1981. *Evolution and Development*. Berlin, Heidelberg, New York: Springer-Verlag.

Britten, R. J., and E. H. Davidson, 1971. Repetitive and nonrepetitive DNA sequences and a speculation on the origins of evolutionary novelty. *Quart. Rev. Biol.* 46:111–137.

Brown, D. D., 1981. Gene expression in eukaryotes. *Science* 211:667–674.

Bush, F. L., 1980. *Essays on Evolution and Speciation*. Cambridge: Cambridge University Press.

Bush, G. L., S. M. Case, A. C. Wilson, and J. L. Patton, 1977. Rapid speciation and chromosomal evolution in mammals. *Proc. Natl. Acad. Sci.* 74:3942–3946.

Cairns, J., 1981. The origin of human cancers. *Nature* (London) 289:353–357.

Charlesworth, B., R. Lande, and M. Slatkin, 1982. A neo-Darwinian commentary on macroevolution. *Evolution* 36:474–498.

Chikaraishi, D. M., S. S. Deeb, and N. Sueoka, 1978. Sequence complexity of nuclear RNAs in adult rat tissues. *Cell* 13:111–120.

Davidson, E. H., and R. J. Britten, 1979. Regulation of gene expression: possible role of repetitive sequences. *Science* 204:1052–1059.

Dickenson, W. J., 1980. Tissue specificity of enzyme expression regulated by diffusable factors: evidence in *Drosophila* hybrids. *Science* 207:995–997.

Dover, G., S. Brown, E. Coen, J. Dallas, T. Strachan, and M. Trick, 1982. The dynamics of genome evolution and species differentiation. In *Genome Evolution*, G. A. Dover and R. B. Flavell, eds. New York: Academic Press, pp. 343–372.

Erdos, P., and A. Renyi, 1959. *On the Random Graphs 1*, vol. 6. Debrecar, Hungary: Inst. Math. Univ. DeBreceniens.

Erdos, P., and A. Renyi, 1960. *On the Evolution of Random Graphs*, publ. no. 5. Math. Inst. Hung. Acad. Sci.

Ewens, W. J., 1979. *Mathematical Population Genetics*. Berlin, Heidelberg, New York: Springer-Verlag.

Flavell, R., 1982. Sequence amplification, deletion and rearrangement: major sources of variation during species divergence. In *Genome Evolution*, G. A. Dover, and R. B. Flavell, eds. New York: Academic Press, pp. 301–324.

Fogelman Soulie, F., E. Goles Chaac, and G. Weisbuch, 1982. Specific roles of the different Boolean mappings in random networks. *Bull. Math. Biol.* 44:715–730.

Gillespie, D., L. Donehower, and D. Strayer, 1982. Evolution of primate DNA organization. In *Genome Evolution*, G. A. Dover and R. B. Flavell, eds. New York: Academic Press, pp. 113–134.

Gould, S. J., and N. Eldredge, 1977. Punctuated equilibria: the tempo and mode of evolution reconsidered. *Paleobiology* 3:115–151.

Gould, S. J., and R. C. Lewontin, 1979. The spandrels of San Marco and the Panglossian paradigm: a critique of the adaptationist programme. *Proc. Roy. Soc.* (ser. B) 205:581–598.

Green, M. M., 1980. Transposable elements in Drosophila and other diptera. *Ann. Rev. Genet.* 14:109–120.

Hastie, N. D., and J. O. Bishop, 1976. The expression of three abundance classes of messenger RNA in mouse tissues. *Cell* 9:761–774.

Hough, B. R., M. J. Smith, R. J. Britten, and E. H. Davidson, 1975. Sequence complexity of heterogeneous nuclear RNA in sea urchin embryos. *Cell* 5:291–299.

Humphreys, T., 1963. Chemical dissolution and *in vitro* reconstruction of sponge cell adhesions, part I: isolation and functional demonstration of the components involved. *Dev. Biol.* 8:27–47.

Hunkapiller, T., H. Huang, L. Hood, and J. H. Campbell, 1982. The impact of modern genetics on evolutionary theory. In *Perspectives on Evolution*, R. Milkman, ed. Sunderland, Massachusetts: Sinauer Assoc., pp. 164–189.

Kauffman, S. A., 1969. Metabolic stability and epigenesis in randomly constructed genetic nets. *J. Theor. Biol.* 22:437–467.

Kauffman, S. A., 1974. The large scale structure and dynamics of gene control circuits: an ensemble approach. *J. Theor. Biol.* 44: 167–190.

Kauffman, S. A., 1982. Developmental constraints: internal factors in evolution. In *British Society for Developmental Biology 6: Development and Evolution*, B. Goodwin and N. Holder, eds. Cambridge: Cambridge University Press, pp. 195–225.

Kauffman, S. A., 1983. Selective adaptation and its limits in automata and evolution. To appear in Proceedings of the Workshop on Dynamical Behaviour of Automata: Theory and Applications. Grenoble: Université Scientifique et Medicale de Grenoble.

Kleene, K. C., and T. Humphrey, 1977. Similarity of hRNA sequences in blastula and pluteus stage sea urchin embryos. *Cell* 12:143–155.

Kurtz, D. T., 1981. Hormonal inducibility of rat globulin genes in transfected mouse cells. *Nature* (London) 291:629–631.

Lewontin, R. C., 1974. *The Genetic Basis of Evolutionary Change*. New York: Columbia University Press.

Moscana, A. A., 1963. Studies on cell aggregation: Decomposition of materials with cell binding activity. *Proc. Natl. Acad. Sci. USA* 49:742–747.

Murray, J. D., 1981. On pattern formation mechanisms for lepidopteran wing patterns and mammalian coat markings. *Phil. Trans. Roy. Soc. London* B295:473–496.

Paigen, K., 1979. In *Physiological Genetics*, J. G. Scandalias, ed. New York: Academic Press, p. 1.

Rachootin, S. P., and K. S. Thompson, 1981. In *Evolution Today*, Proceedings of the Second International Congress on Systematic and Evolutionary Biology, G. C. Scudder and J. L. Reveal, eds. Pittsburgh: Carnegie-Mellon University Press.

Renel, J. M., 1967. *Canalisation and Gene Control*. New York: Academic Press.

Roeder, G. S., and G. R. Fink, 1980. DNA rearrangements associated with a transposable element in yeast. *Cell* 21:239–249.

Schopf, T. J. M., 1980. In *Paleobotany, Paleo-ecology and Evolution*, K. J. Niklas, ed. New York: Praeger.

Shapiro, R. A., 1981. Changes in gene order and gene expression. *Nat. Cancer Inst. Monographs*. In press.

Simpson, G. G., 1943. *Tempo and Mode in Evolution*. New York: Columbia University Press.

Smith, J. M., 1982. Overview—Unsolved evolutionary problems. In *Genome Evolution*, G. A. Dover and R. B. Flavell, eds. New York: Academic Press.

Thomas, R., 1979. *Kinetic Logic: A Boolean Approach to the Analysis of Complex Regulatory Systems*. Lecture Notes in Biomathematics 29. New York: Springer-Verlag.

Waddington, C. H., 1975. *The Evolution of an Evolutionist*. New York: Cornell University Press.

Wake, D. B., G. Roth, and M. H. Wake, 1983. On the problem of stasis in organismal evolution. *J. Theor. Biol.* 101:211–224.

Webster, C., and B. C. Goodwin, 1982. The origin of species: a structuralist approach. *J. Soc. and Biol. Structure* 5:15–42.

Willmer, E. N., 1960. *Cytology and Evolution*. New York: Academic Press.

Wilson, A. C., T. J. White, S. S. Carlson, and L. M. Cherry, 1977. In *Molecular Human Genetics*, R. R. Sparks and D. E. Comings, eds. New York: Academic Press.

Winfree, A. T., 1980. *The Geometry of Biological Time. Biomathematics*, vol. 8. New York: Springer-Verlag.

Hermeneutics and the Analysis of Complex Biological Systems

Gunther S. Stent

Practicing scientists are wont to regard the philosophy of science as an activity of has-beens or parasitic ne'er-do-wells, a subject suitable at best for discussion over brandy after dinner. That view is not wholly unjustified, since many lines of scientific work can be successfully pursued without any clear understanding of their philosophical basis. For instance, as I know from my own personal experience, most of the founding fathers of molecular biology (with such notable exceptions as Max Delbrück and François Jacob) managed to revolutionize twentieth-century biology without taking any interest in philosophy. That is not to say, of course, that scientists are not informed by the philosophical ideas of their surroundings or times, which they clearly are, as is almost everyone else. But these ideas usually enter into scientific work only implicitly, by way of notions held by the scientist tacitly, without any clear, or even conscious, awareness of their content. Nonetheless, in the history of science, every once in a while, there have been cases in which explicit philosophical considerations did play a crucial role in a fundamental scientific advance. For instance in the development of relativity theory and of quantum mechanics some necessary philosophical groundwork was done, not by professional philosophers but by the scientists themselves, Einstein and Niels Bohr.

It is my belief that current efforts to analyze and understand complex biological systems happen to present yet another of those rare cases where some philosophical attention (or "intellectual hygiene," as my teacher André Lwoff called it) would be of some benefit to further scientific progress. Here I am thinking not only of such obvious examples of biological complexity as ecosystems or evolution, whose intrinsic epistemological difficulties are quite generally appreciated, but I am thinking also of such less obvious examples as the vertebrate immune and nervous systems, whose study seems to be squarely within the domain of hard-nosed experimental biology. In this chapter I will focus my attention on the nervous system, since it is the system I am studying. The nervous system consists of a complex network of very many specifically interconnected nerve cells, or neurons, which govern animal behavior. So the nervous system poses two deep problems for the biologist. First, how does this complex network manage to function and do what it does do? And how does this complex network, with its myriad of specifically interconnected cells, manage to arise during the embryological development of the animal?

Programmatic Phenomena

The goals of developmental neurobiology are currently in want of conceptual clarification, precisely because of the tremendous explanatory power that the recent insights of molecular biology, and of its central concept of biological information, have brought to a broad range of biological problems. But in my view, at least, these molecular biological insights are now being misapplied to the problems of embryonic development in general, and to developmental neurobiology in particular. My criticism of these misapplications of molecular biology is not that they are reductionist—I think the time has come to ban the use of that debauched epithet from polite society—but that they are misapplications. As for developmental neurobiology, one such misapplication arose in the mid–1960s, at the time of the triumphant culmination of molecular biological research in the cracking of the genetic code. At that time research projects began to be formulated that attempted to combine the disciplines of genetics and of developmental neurobiology. These projects were evidently inspired by the idea that the deep biological problem of how the cellular components and precise interconnections of the nervous system arise during development could, and even should, be approached by focusing on genes. In particular, the notion arose that the structure and function of the nervous system, and hence the behavior of an animal, is specified by its genes. Some of my fellow molecular biologists even went so far as to propose that the genes actually contain the circuit diagram of the nervous system. It cannot be the case, of course, that the genes really embody enough information to permit explicit specification of a neuron-by-neuron circuit diagram (Brindley, 1969). And so a seemingly more reasonable view of the nature of the genetic specification of the nervous system would be that the genes embody, not a circuit diagram, but a program for the development of the nervous system as was once proposed by Sydney Brenner (1973, 1974). Accordingly, the main goal of developmental neurobiology would be to discover that genetic program. But this view of the development of the nervous system is rooted in a semantic confusion about the concept of program. Once that confusion is cleared up, it becomes evident that development of the nervous system is unlikely to be a programmatic phenomenon. And so I would like to explicate first the nature of programmatic phenomena.

Development belongs to the class of regular phenomena which share the property that a particular initial situation generally leads, via a more or less invariant sequence of intermediate steps, to a particular final situation. However, within the class of regular phenomena, programmatic phenomena form only a small subclass, almost all the members of which are associated with human activity. For a phenomenon to belong to the subclass of programmatic phenomena it is a necessary condition that, in addition to the phenomenon itself, there exist a second thing, namely, the program. And the structure of that second thing is isomorphic with, that is, can be brought into one-to-one correspondence with, the phenomenon of which it is the program. For instance, the on-stage events associated with a performance of *Hamlet*, which is a regular phenomenon, are programmatic since there exists Shakespeare's text with which the on-stage actions of the performers are isomorphic. But the no less regular off-stage events, such as the actions of the house staff and audience, are mainly nonprogrammatic, since their regularity is merely the automatic consequence of the contextual situation of the performance. To give another example, the operation of a digital computer has programmatic aspects, insofar as there exists a program, or set of instructions separate from the hardware, whose structure is isomorphic with the sequence of operations performed by the machine.

One of the very few regular phenomena independent of human activity that can be said to have a programmatic component is the formation of proteins. Here the assembly of amino acids into a polypeptide chain of a given primary structure *is* programmatic because there exists a stretch of DNA polynucleotide chain—the gene—whose nucleotide base sequence is isomorphic with the sequence of events that unfolds at the ribosomal polypeptide assembly site. However, the subsequent folding of the completed polypeptide chain into its specific tertiary structure lacks programmatic character, since the three-dimensional conformation of the molecule is the automatic consequence of its contextual situation and has no isomorphic correspondent in the

DNA. This example of the formation of proteins can serve also to clarify the distinction I made earlier between the embodiment by the genes of a neuron-by-neuron circuit diagram of the nervous system on the one hand, and a program for its development on the other. Evidently in the case of proteins, the genes do not embody an explicit atom-by-atom specification of spatial coordinates of the tertiary structure of proteins, i.e., a circuit diagram, but merely a program for assembly of their primary structure from ready-made amino acid building blocks.

When we extend these considerations to the regular phenomenon of development, we see that its programmatic aspect is confined mainly to the assembly of polypeptide chains (and of various species of RNA). But as for the overall phenomenon, it is most unlikely—and no credible hypothesis has as yet been advanced to explain how this could be the case—that the sequence of developmental events is isomorphic with the structure of any second thing, especially not with the structure of the genome. The fact that mutation of a gene leads to an altered phenotype shows that genes are part of the causal antecedents of the adult animal, but does not in any way indicate that the mutant gene is part of a program for its development.

But are not polemics about the meaning of words such as *program* just a waste of time for those who want to get on with the job of finding out how the nervous system really develops? As J. H. Woodger (1952) showed in his Tarner Lectures entitled *Biology and Language,* published shortly before Watson and Crick's discovery of the DNA double helix and Seymour Benzer's reform of the gene concept, semantic confusion about its fundamental terms, such as 'gene', 'genotype', and 'determination' had become the bane of classical genetics. It would be well to avoid reinstituting that confusion in the context of developmental biology and to remember Woodger's advice that "an understanding of the pitfalls to which a too naive use of language exposes us is as necessary as some understanding of the artifacts which accompany the use of microscopical techniques" (p. 6).

The general notion of genetic specification of the nervous system is defective not only at the conceptual level but also represents a misinterpretation of the knowledge already available from developmental studies. As George Székely (1979) has pointed out, we know enough about neuronal development already to make it most unlikely that neuronal circuitry is, in fact, prespecified; rather, all indications point to stochastic processes as underlying the apparent regularity of neural development. That is to say, development of the nervous system, from fertilized egg to mature brain, is not a programmatic but a historical phenomenon under which one thing simply leads to another. To illustrate the difference between programmatic specification and stochastic history as alternative accounts of regular phenomena, we may consider the immune response, or the establishment of ecological communities upon colonization of islands (Simberloff, 1974), or the growth of secondary forests (Whittaker, 1970). All three of these examples are regular phenomena. In the case of the immune response the introduction of an antigen into the circulation of an animal regularly elicits production of antibody molecules having a high specific affinity for that antigen. These antibody molecules arise via a progressive and predictable change in the genetic character of the animal's lymphocyte population. And in the cases of island colonization and secondary forest growth a more or less predictable ecological structure arises via a stereotyped pattern of intermediate steps, in which the relative abundances of various types of plants and animals follow a well-defined sequence. But the regularity of these phenomena is obviously not the consequence of an immunological or of an ecological program encoded in the genome of the immunized animal or of the colonizing species. Rather, in all three examples the regularity is a consequence of a historical cascade of complex stochastic interactions between various cell types or between various creatures and the world as it is. It appears therefore that rather than uncovering a gene-encoded program, the goal of developmental neurobiology must be the discovery of the functional relations, or algorithms, that govern the nonprogrammatic, contextually determined intra- and intercellular interactions underlying the historical phenomenon of metazoan ontogeny.

Hermeneutics

As most historians, but not so many scientists, know, to fathom the complex interactions that produce historical phenomena, it is necessary to understand the context in which they are embedded. Appreciating the importance of contextual relations for the problem of embryonic development, we are led into a domain of phenomenological analysis to which the traditional views of scientific research are no longer fully applicable. That domain bears a strong epistemological affinity to the scholarly activity called hermeneutics. This designation was originally given by theologians to the methodology of interpretation of sacred texts, especially of the Bible. The name is derived from that of Hermes, the divine messenger. In his capacity as an information channel linking gods and men, Hermes must interpret, or make explicit in terms that ordinary mortals can understand, the implicit meaning that is hidden in the gods' messages. In recent years, scholars have extended the term *hermeneutics* to the interpretation of secular texts, since there may be implicit meanings hidden even in the literary creations of ordinary men that need to be made explicit to their fellow mortals. But hidden meanings pose a procedural difficulty for textual interpretation, because one must understand the context in which implicit meaning is embedded before one can uncover hidden meanings in any of its parts. Here we face a logical dilemma, the so-called hermeneutic circle. On the one hand, the words and sentences of which a text is composed have no meaning until one knows the meaning of the text as a whole. On the other hand, one can only come to know the meaning of the whole text through understanding its parts. To break this circle hermeneutics invokes the doctrine of pre-understanding. As set forth by Rudolf Bultmann and Martin Heidegger, hermeneutic pre-understanding, or *Vorverständnis*, represents the life of experience and insights that the interpreter must bring to the task of interpreting.

In assessing the epistemological status of hermeneutic studies we may ask to what extent the concept of objective validity is applicable to their results. According to the traditional view of hermeneutics, an objectively valid interpretation is one that has made explicit the true meaning hidden in the text, that is, the meaning intended by the author. But here we encounter two difficulties. First, the author may not have been—in fact, according to the teachings of analytical psychology, most likely was not—consciously aware of the (subconsciously) intended meaning of his own text. Therefore the outcome of the only operational test of the validity of an interpretation, namely, asking the author: Is this your intended meaning? (or discovering the author's own explicit statement of the meaning of his text), does not provide an objective criterion of interpretative truth. What would be needed in addition is an (also interpretative) exploration of the author's subconscious. Second, in order to be eligible for even attempting a true interpretation in the first place, the interpreter must possess just those experiences and insights that the author presupposed (consciously or subconsciously) in the audience to which his text is addressed. But those experiences and insights, that is, the interpreter's pre-understanding, are necessarily based on his own subjective historical, social, and personal background. Hence agreement regarding the validity of an interpretation could be reached only among persons who happen to bring the same pre-understanding to the text. Thus because of the conceptual lack of an operational test for truth, on the one hand, and the necessarily subjective nature of pre-understanding on the other, there cannot be such a thing as an objectively valid interpretation. Contemporary students of hermeneutics, such as Hans-Georg Gadamer, even assert that the very concept of the true meaning is incoherent, inasmuch as they claim that a text can have very many meanings, of which the author's intended meaning is only one, and not even a privileged one. Thus it is this evident unattainability of universal and eternal truth in interpretation that seems to make hermeneutics different from the Greek conception of science, for which the belief in the attainability of objectively valid explanations of the world is metaphysical bedrock.

To what extent is this traditional belief in objectively valid explanations in science actually justified? According to some contemporary philosophers of science, such as Thomas Kuhn and Paul Feyerabend, it is not justified because the scientist also must bring subjective notions equivalent to hermeneutic pre-understanding to his search for explanations of phenomena. Indeed, as was pointed out by Ludwik Flek (1979), in the early 1930s the still unappreciated precursor of these latter-day radical critics of eternal truth of scientific explanation, even the so-called facts of science, are not objective givens but socially conditioned creations. Nevertheless, even if we admit this radical critique of the concept of objective validity in scientific explanation, it nevertheless seems that the analysis of some phenomena requires less pre-understanding than of others, and hence that some explanations might be able to lay claim to a relatively closer approach to objective validity than others. Thus we could estimate the degree to which a scientific explanation might be objectively valid by assessing the extent to which pre-understanding played a role in its development. Such an assessment can help us to understand why the Greek belief in the attainability of objectively valid explanations is somewhat more justified in the hard natural sciences, such as physics, than in the soft human sciences, such as economics, sociology, and psychology.

One of the main reasons for this difference in the degree to which claim can be laid to objective validity is that the phenomena which the soft sciences seek to explain are much more complex than those addressed by the hard sciences. What do we actually mean by the complexity of a phenomenon? Here it is important to appreciate that the complexity of a phenomenon is not to be measured by the number of component events of which it is constituted, but rather by the diversity of the interactions among its component events. For instance, according to that criterion, a 1 gallon bottle containing 10^{23} molecules of hydrogen gas is a very simple phenomenon, since despite their tremendous number the molecules interact in only one way: they collide and exchange momentum. Hence by taking just that one interaction into account it is possible to give a highly satisfactory statistical mechanical explanation of the pressure exerted by the gas on the walls of the bottle. By contrast, a city of 10^6 inhabitants is a highly complex phenomenon, since these inhabitants interact in a myriad of different ways. Hence to develop a sociological explanation of, say, the incidence of murder in such a city, it is clearly impossible to take into account more than a tiny fraction of these interactions, namely, those which the sociologist deems relevant to the problem at hand. It follows, therefore, that the more complex the ensemble of events that the scientist isolates conceptually for his attention, the more hermeneutic pre-understanding must he bring to the phenomenon before he can break it down into meaningful atomic components that are to be governed by the causal connections of his eventual explanations. Accordingly, the less likely is it that his explanations will have the aura of objective validity.

By way of comparing a pair of extreme examples—one very hard, the other very soft—we may consider mechanics and psychoanalysis. There is an aura of objective validity about the laws of classical mechanics because the phenomena which mechanics consider significant, such as steel balls rolling down inclines, are of low complexity. Because of that low complexity, it is possible to dissect a phenomenon into the essential components—steel ball and incline—whose interaction is governed by the causal connections envisaged by the explanation, without having to invoke very much pre-understanding. To validate the explanation one can adduce critical observations or experiments with various kinds of steel balls and inclines. By contrast, there is no comparable aura of objective validity about the propositions of analytical psychology, because the phenomena of the human psyche which it attends are very complex. Without resort to far-reaching pre-understanding, no analyst can recognize any structure in, let alone dissect, the phenomenon (the psyche of his analysand) into its essential, causally connected components. In psychoanalysis there are no critical observations or experiments, because the failure of any prediction based on psychoanalytic theory can almost always be explained away retrodictively by modifying slightly one's pre-understanding of the phenomenon. Hence in psychoanalysis a counterfactual prediction rarely qualifies as negative evidence against the theory that generated it, which is why many scientists deny standing—in my opinion unjustifiably— to psychoanalysis as a scientific discipline. Sigmund Freud himself failed to appreciate this fundamental epistemological limitation of psychoanalysis. He thought he had founded the physics of the mind. But as it turned out, he had founded its hermeneutics. As pointed out by Jurgen Habermas (1968), psychoanalysis consists of the hermeneutic interpretation of the complex text that is provided to the analyst by his subject. This misunderstanding by Freud still remains at the root of the ambiguous relation of psychoanalysis to the natural sciences.

Neural Networks

I now turn to the other deep problem posed by the nervous system, namely, how its complex network manages to function. As we can now appreciate, neurobiology spans a broad range on the hardness-softness scale of the sciences. At its hard end, neurobiology is represented by electrophysiological, anatomical, and biochemical studies of nerve cells. Although the phenomena associated with nerve cells are more complex than rolling steel balls, they can still be accounted for by explanations that are susceptible to seemingly objective validation. But at its soft, and to me more fascinating, end, neurobiology is represented by system-analytical studies of the structure and function of large and complicated cellular networks. The phenomena associated with neural networks approach the human psyche in their complexity; in fact, they include the human psyche.

As I have found from experience, hermeneutic pre-understanding is required even for the functional analysis of nerve cell networks that seem ridiculously simple in comparison with those we encounter in the human brain. About twelve years ago I embarked on a research project whose objective was to discover the nerve cell network in the leech that is responsible for generating the wave-like swimming rhythm of that bloodsucking worm. The project I had chosen seemed feasible because, in contrast to the billions of nerve cells that make up the brain and spinal cord of even the lowliest vertebrate animals, the central nervous system of the invertebrate leech consists of a chain of 32 iterated nerve cell aggregates, or segmental ganglia, each of which contains more or less the same complement of only about 200 bilateral pairs of identifiable nerve cells. Now although the total number of different ways in which 200 pairs of neurons might interact with one another via their connections is still astronomically large, it did seem at least conceivable that, with some luck, we might be able to run the leech nervous system into the ground and find the swim generator. After about six years of hard work, my associates and I had managed to identify an ensemble of about a dozen pairs of swim motor neurons in each ganglion of the leech nervous system that make connections with the muscles responsible for executing the swimming movement and whose rhythmic activity appears to command the rhythmic contraction-distension pattern of those muscles needed for swimming. Moreover, we managed to find four pairs of swim oscillator neurons in each ganglion which not only are interconnected in such a manner that their network should generate the periodic impulse pattern of the swim rhythm but also are connected to the swim motor neurons in a manner appropriate for imposing that rhythm on them. To convince ourselves that this neuronal network actually did possess the rhythm generator capacity we attributed to it, we built an electronic analog circuit in which the electrophysiological properties of the identified neurons and their connections were imitated. To our great satisfaction the activity of the electronic analog circuit mimicked very closely the activity pattern of the real swimming rhythm (Stent et al., 1978).

So we were (and we still are) convinced that we had found at least the principal neuronal elements of the leech swim generator network. But when I proudly presented our findings at neurobiological meetings, I was often asked by colleagues, who were similarly in search of neuronal networks responsible for other rhythmic movements, whether my circuit diagram of the alleged leech swim generator network actually contained all the neurons which we had found to display an activity rhythm during swimming and all the connections we had found to be made by such neurons. To this question I had to reply in the negative: no, in fact, we had found more rhythmically active neurons and more interconnections than those shown on the circuit diagram. Then why hadn't I shown them? Because, I answered, according to George Székely's theory of recurrent cyclic inhibition oscillator networks, which we used to explain the function of the swim generator, they were not needed. This reply usually evoked laughter in the hall and the comment by our critics that we were obviously not serious neurobiologists, since we only got out of our studies what we had projected into them. Unlike us, serious students of neuronal networks do not select only those of their findings that happen to fit some theoretical preconception. Rather they take all their data into account and then model them on a digital computer, to make a strictly empirical, honest examination of what the identified network might actually be doing.

The work of our critics may be more honest than ours, but their so-called empirical procedure—which is not, and of course cannot be, entirely theory-free either—is most unlikely to lead to the identification of the neuronal circuits underlying even rather simple behavioral routines. Since the topology of the circuitry of any nervous system, be it ever so small, is so highly interconnected that there is usually some neural or hormonal pathway leading from any point to any other point, it is simply not possible to start any functional analysis without having some prior theoretical insight into the problem. Hence at its soft end neurobiology takes on some of the characteristics of hermeneutics: the student of a complex neural network must bring considerable pre-understanding to the system as a whole before attempting to interpret the function of any of its parts. Accordingly, the explanations that are advanced about complex neural systems may remain beyond the reach of objective validation.

Visual Perception

In fact, not only must the would-be analyst of the nervous system bring conscious pre-understanding to his or her task, but must also depend on his or her own nervous system to interpret preconsciously, sensory experience of the outside world according to hermeneutic principles. This insight has emerged slowly during the past twenty years from the study of the process of visual perception, especially thanks to the work of the recently deceased young English mathematician, David Marr (1982). Visual perception can be regarded as a hermeneutic activity, by means of which the viewer interprets the meaning that is hidden in a visual scene. The process begins with an image cast by the visual surround upon the retina and its mosaic of light receptor cells. That image may be considered a matrix of numerous elements, each characterized by a given level of light intensity. This matrix is called the *gray-level array*. As analyzed by Marr, visual perception culminates in a meaningful description of the image, such as This is an animal, or This is a bear, or This is the head of a teddy bear. Evidently, this description depends not only on the image itself but also on the context in which it is produced and on the purpose that the viewer brings to it. In other words, Marr approaches perception as a process that makes explicit for the viewer the meaning of the information that originally is merely implicit in the image. In this regard, Marr's approach has strong links to present-day cognitive psychology, in general, and to linguistics, in particular. For these disciplines, the extraction of explicit meaning from information in which that meaning is merely implicit—that is to say, the performance of semantic decoding—is the central (and as yet unsolved) problem.

The focus of Marr's theory of vision is on the perception of shape. Marr is not the first, by any means, to view the problem of visual perception in these semantic terms. But his approach differs from that of his predecessors, who sought to interpret the image by means of a search process that scans the retinal image for the presence of regions that are meaningful as specific objects. For instance, according to these earlier theories, a hammer would be identified in a retinal image by tentatively labeling a dark blob as a hammerhead and then looking in the vicinity of the blob for a form that fits the description of a hammer handle, which (as the viewer knows) normally ought to be attached to a hammerhead.

If this approach to visual perception were correct, then just the right pieces of specialized knowledge, or pre-understanding, would have to be made available from a vast cerebral library at the very earliest stages of the image-interpretation process. Marr, by contrast, aims at a process that squeezes the last possible bit of meaning from the image before calling on any pre-understanding stored in higher brain centers about specific objects. Instead of requiring such specific pre-understanding as that hammerheads are normally attached to handles, Marr's theories demand at early stages of visual processing only general knowledge about the world, such as that it is constituted mainly of solid, non-deformable objects of which only one can occupy a given place at any time. This much more limited general knowledge can then be incorporated as a fixed part of early image processing. In fact, such general knowledge can be taken into account in the design of the retina and hence need not descend from higher brain centers for the initial stages of the semantic decoding of the image. Marr's invocation of general rather than specific pre-understanding of the world is one of the most distinctive features of his theories.

The starting point of Marr's theory of image processing lies in the commonplace experience that a real scene and an artist's sketch of that scene evoke similar percepts in the viewer, despite the fact that they give rise to very different retinal images. This fact suggests that the artist's sketch corresponds in some way to an intermediate stage of the process by which the percept is extracted from the image of the real scene. Accordingly, Marr envisages that perception begins with a transformation of the image into what he calls the *primal sketch*.

The idea underlying both design and interpretation of the primal sketch is the pre-understanding that contour outlines, and hence the shapes, of objects in the visual surround are represented by those domains in the image where there occurs an abrupt change in light intensity. So, to make explicit the presence of contour outlines, which are merely implicit in the image, it is first necessary to describe the way in which the light intensity changes from place to place in the image. This description is the primal sketch.

In order to generate the primal sketch, the image has to be subjected to just the kind of differential analysis that is known to be carried out by the neurons discovered by David Hubel and Torsten Wiesel in the visual cortex. Marr assigns to these neurons the function of measuring the rate of change of light intensity along a given direction. Hence, by simultaneous scanning of the entire image, the ensemble of these cortical neurons extracts from the image the overall pattern of spatial variation in light intensity. In the primal sketch, which can be extracted from the retinal image by scanning it to an algorithm designed by Marr, each line represents the position and orientation of a spatial change in intensity in the image. The primal sketch thus makes explicit the positions, directions, magnitudes, and spatial extents of light intensity gradients present in the image.

The goal of the next stage in the semantic decoding process is to determine which of the lines of the primal sketch correspond to contour outlines of objects, and which lines merely correspond to continuous changes in surface orientation relative to the viewer. Or, in perceptual terms, the problem that must now be solved is the separation of figure from ground. Various visual processes can be called on for this purpose. One of these processes is stereopsis, or the decoding of depth information from the two slightly different images that an object produces in the right and left eyes. As Marr could show, to make this decoding the viewer need not resort to specific pre-understanding of familiar shapes and forms. Instead, it suffices that his visual system takes into account two commonplaces of general pre-understanding regarding the nature of the world: physical objects can be in only one place at a time, and they are cohesive. In addition to stereopsis, surface shape can also be decoded from the primal sketch by processes that operate on information provided by shading, texture gradients, and perspective cues present in the image, as well as by motion. As the 3-D impression of TV-screen displays of rotating molecular models demonstrate so dramatically, the change in appearance of a moving object provides information regarding its shape. Marr's associate, Shimon Ullman, was able to solve the problem of how the shape of an object can be derived from its moving image by positing the general pre-understanding that the object in motion is rigid and has not changed its shape in the time elapsed between consecutive images. For such a rigid object Ullman proved a theorem which states that three distinct views of four noncoplanar points are sufficient to determine uniquely their arrangement in space.

To combine the information about surface shape provided by these various processes in some manner suitable for further processing, a description is needed that makes explicit the surface contours. Marr and his colleague, H. Keith Nishihara, call this description the 2-1/2 dimensional, or *2-1/2-D sketch*. In that sketch dotted lines indicate surface contours and the orientation (relative to the viewer) of small patches of surface spaced evenly over the entire image is represented by arrows. With this explicit description of the surface contours implicit in an image, figure has been separated from ground, without resort to higher level specific pre-understanding of the geometrical shapes of objects.

The next stage of the processing of visual information must concern the conversion of the 2-1/2-D sketch into a description of 3-dimensional shapes suitable for object recognition. That is to say, this description must confer on the viewer the ability to recognize a shape as being the same as a shape seen earlier. And this ability depends on being able to describe a shape consistently each time it is seen, regardless of its position relative to the viewer. For the purpose of such a description, Marr proposes that the viewer reduces the object to an ensemble of natural components, so as to allow him to recognize its shape from a description of the relative spatial arrangement of these components in terms of their location, sizes, and principal axes.

For instance, Marr presented such a description of a human shape in terms of a hierarchy of sticks that he calls the *3-D model*. To construct the 3-D model, the overall form—the body—is first assigned an axis. This assignment allows the establishment of an object-centered (rather than viewer-centered) coordinate system which can be used to describe the arrangement of the arms, legs, torso, and head. The position of each of these components is described by the relative orientation of its own axis, which in turn serves to define a secondary coordinate system for describing the arrangement of further subsidiary components. These successive axial assignments generate the hierarchy of 3-D models, shown here as extending down to the level of fingers. Under this hierarchical scheme, any component of a shape can be treated as a shape in itself and thus the final description of the shape can be carried down to any arbitrary level of detail.

For such a description of a shape in terms of a 3-D model to be recognizable, it is necessary that the mode of decomposition into axial subcomponents of the shapes in the 2-1/2-D sketch is more or less independent of the conditions of observation that generated the image. To achieve this end, the choice of the object-centered coordinate system must be governed by some salient geometric characteristics possessed by the shape. For instance, the coordinate system for a sausage should take advantage of its major axis, and that for a human face, of its axis of symmetry. What sort of shapes have an axis as an integral part of their structure? The answer is: all objects belonging to the class of generalized cones. What is a generalized cone? It is the surface swept out by moving closed contours of constant shape but smoothly varying size along an axis. Fortunately, many objects, especially those whose shape was generated by growth, are describable quite naturally in terms of one or more generalized cones.

But how is the coordinate frame of a shape composed of generalized cones to be found from the surface contours of the 2-1/2-D sketch for construction of the 3-D model? Here Marr puts forward three assumptions that the viewer makes intuitively about the way a surface contour is generated. First, each point on the contour is projected from a different point on the object. Second, nearby points on the contour are projected from nearby points on the object. And third, all points on the contour are projected from points lying in a single plane. Using such pre-understanding, Marr posited a process by which the coordinate frame can be extracted from the surface contours of the retinal image of a shape composed of generalized cones.

To complete the process of perception, or of making explicit the meaning of the image, the shape of the 3-D model must be recognized as a particular object. It is at that stage that the viewer's specific, high-level pre-understanding must finally be called on. According to Marr, that high-level pre-understanding consists of a catalog of stored 3-D models. This catalog is indexed in various ways, so that a 3-D model can be identified with a stored catalog item. A 3-D model extracted from an image may thus be identified with a catalog item. Upon the identification of the 3-D model with a catalog item, the goal set by Marr of reaching a meaningful description of the image has been reached.

At this point, a tough-minded experimentally oriented neurobiologist might ask whether there is any chance of ever proving such a theory as Marr's, or whether it is just an idle speculation, only suitable, like philosophy, for discussion over brandy after dinner. To deal with such skepticism, we return to the general subject of hermeneutics. For Marr not only explains visual perception as a hermeneutic process resorting to preconscious pre-understanding but also bases his very explanation of that process on an interpretative procedure heavily dependent on his own (conscious) pre-understanding of how the brain probably does its work. Hence his theory is likely to remain beyond the reach of objective validation.

Looking therefore at Marr's approach to vision from the much less tough-minded, but more realistic hermeneutic perspective, one cannot but admire its attempt at rigor in the analysis of what, on first sight, appears to be a hopelessly complex biological system. Marr's conceptual identification at each level of the hierarchical perception process of the minimum general pre-understanding that must be invoked has an aura of validity. Most important, Marr's theory, however incomplete for the time being, has brought a new way of thinking about visual perception. Marr's theory may never be completely connectable to neurophysiology; and maybe it isn't even true. But for the time being, at the soft end of neurobiology, it's the best we've got. Or to paraphrase my former employer, California Governor Edmund G. Brown, Jr., in the analysis of complex biological systems we may have to be satisfied with less.

References

Brenner, S., 1973. The genetics of behaviour. *Brit. Med. Bull.* 29:269–271.

Brenner, S., 1974. The genetics of *Caenorhabditis elegans*. *Genetics* 77:71–94.

Brindley, G. S., 1969. Nerve net models of plausible size that perform many simple learning tasks. *Proc. Roy. Soc. London B* 174:173–191.

Flek, L., 1979. *Genesis and development of a scientific fact*. T. J. Trenn and R. K. Merton, eds.; F. Bradley and T. J. Trenn, transl. Chicago: University of Chicago Press.

Habermas, J., 1968. *Erkenntnis und Interesse*. Frankfurt a.M.: Suhrkamp.

Marr, D., 1982. *Vision*. San Francisco: W. H. Freeman.

Simberloff, D. S., 1974. Equilibrium theory of island biography and ecology. *Ann. Rev. Ecology Systematus* 5:161–182.

Stent, G. S., W. B. Kristan, Jr., O. W. Friessen, C. A. Ort, M. Poon, and R. L. Calabrese, 1978. Neuronal generation of the leech swimming movement. An oscillatory network of neurons driving a locomotory rhythm has been identified. *Science* 200:1348–1357.

Szekely, G., 1979. Order and plasticity in the nervous system. *Trends in Neurosci.* October:245–248.

Whittaker, R. H., 1970. *Communities and Ecosystems* New York: MacMillan.

Woodger, J. H., 1952. *Biology and Language*. Cambridge: Cambridge University Press.

Innovation and Tradition in Evolutionary Theory:
An Interpretive Afterword

David J. Depew
and
Bruce H. Weber

We suggest that the transformation of evolutionary theory envisaged in this volume can best be understood in terms of the philosophy of science advocated by several of the contributors. This chapter is thus a meditation on the essays presented here and an invitation to reflection on the part of the reader. It is important to note that the authors whose work we comment on may or may not agree with our observations.

The prospective transformation of evolutionary theory has at least two aspects. First, it embraces (in the face of well-reasoned dissent by Francisco Ayala) an "expansion" of neo-Darwinism to recognize more fully the hierarchical structure of biological systems and the possibility that selection occurs at any number of these hierarchical levels. Second, it acknowledges the need for a shift away from emphasis on selection alone toward a consideration of other evolutionary processes.

Such a transformation might well portend a breakdown of the neo-Darwinian synthesis if that synthesis is strongly wedded to received empiricist conceptions of science. As we argue in the first and second sections, reductionistic ideals, which form an integral part of doctrinaire empiricism, do not accord well with the hierarchical structure of biological systems. In the third section we propose that certain themes articulated by new or post-Kuhnian philosophers of science provide a far less problematic background against which an expansion of neo-Darwinism can be viewed. In the final section we suggest that the same perspective will be helpful in judging how far nonselectionist accounts of evolution are complementary to, and integrable with, neo-Darwinian thought.

Reductionistic Ideals and Hierarchical Realities

Since the seventeenth century scientists have claimed with growing self-confidence to be in possession of a method of gathering and assessing information radically more reliable than those characteristic of other human pursuits. Empiricist proponents of this view have held that by breaking experience down into least units of perception, thus freeing them from habitual interpretive frameworks, scientists are in a position systematically to re-integrate these data into objective conceptual patterns. On the basis of this belief there grew up an influential view that only judgments reached by methods peculiar to empirical science could count as knowledge, and that only these judgments could be aggregated over time in a way that is securely cumulative and progressive, in contrast to the hurly-burly of our religious, cultural, and political beliefs.

The fullest defense of this idea—the logical empiricism of our own century—has honored the inductivist demands inherent in it by putting much stress on the role of predictive tests in confirming or disconfirming scientific hypotheses.Predictive success was presumed to lead to the confirmation of hypotheses and predictive failure to their falsification. In this way a certain picture of scientific progress took shape. The testing process would continually expand the relevant data base, while setting afoot a search for ever better theoretical structures. In the end, nothing would remain in an initially hypothetical theoretical structure that was not strictly justified by the data, while all the data would be logically deducible from a well-confirmed theoretical structure.

An integral aspect of modern empiricism is its conviction that continued application of these methods would tend to show over time that propositions provisionally regarded as central to one science could be deductively inferred from, and hence reduced to, the principles of a more basic one. To make the higher depend on the lower in this fashion offers progressively greater assurance that a priori interpretive suppositions transcending the given are being steadily eliminated. In this way theory-reduction is connected to epistemological reduction.

The neo-Darwinian synthesis was articulated and diffused within a philosophical and cultural climate deeply affected by this program. Thus even while neo-Darwinists resisted any premature reduction of evolutionary biology to physics or chemistry, they were eager to show that evolution at and above the species level could be accounted for in terms of microevolutionary changes in gene frequencies observable and inducible in the laboratory. They were also generally pleased to receive congratulations from empiricist philosophers of science for pushing the unity-of-science program along the reductionist road (Ghiselin, 1969; Ruse, 1973, 1982).

This subtle interaction between neo-Darwinism and reductionistic ideals may be seen to good effect in the recent debate about the "punctuated equilibrium" model of macroevolution proposed by Stephen Jay Gould and others (Eldredge and Gould, 1972; Gould and Eldredge, 1977; Stanley, 1979). Our concern is less with the adequacy of this model than with the pattern of discussion it has evoked.

The punctuated equilibrium model advances the idea that the branching of lineages cannot be accounted for solely in terms of the processes of anagenetic change observed in the laboratory by population geneticists. More positively, the model asserts that organic diversification is concentrated at the branching points where speciation events occur. Punctuated equilibrium allows paleontologists to regard gaps in the fossil record as evidence rather than as the lack of evidence. Moreover, by implying that lineages undergo sudden change followed by long periods of stability, it reaffirms the sympathy paleontologists have continued to feel with continental traditions in morphology. These tend to view evolutionary change as the reorganization of basic morphological patterns (*Baupläne*), rather than as the product of selective forces working independently on various organic parts (Grene, 1983; Gould and Lewontin, 1979).

There has been a tendency to greet the punctuated equilibrium model as a possible refutation of neo-Darwinism. However, the revelation that nature sometimes exhibits patterns projected by punctuated equilibrium can count as a refutation of neo-Darwinism only when the latter is construed as a claim to have reduced macroevolution to microevolution. In that case, one instance of punctuated equilibrium might be taken as disconfirming neo-Darwinism. The existence of this sort of talk is itself testimony to the entanglement of the synthesis in reductionistic rhetoric.

It is against this background that we may most profitably view Francisco Ayala's defense, in this book and elsewhere (Stebbins and Ayala, 1981), of the macroevolutionary work of the synthesists. Ayala is anxious to show that whatever long-range hopes for reduction neo-Darwinists might entertain, they do not have to claim to have achieved a reduction of macroevolution to microevolution, and so do not have to open themselves to attack based on the possible applicability of the punctuated equilibrium, or any other, model. For Ayala macroevolution remains a distinct field to which various models may or may not be relevant. He suggests that the significance of the punctuated model may have been distorted by failure on the part of its proponents to realize that what counts in paleontology as a geological instant is time enough for a population geneticist's gradualism.

Stephen Jay Gould, for his part, has not been among those who take the punctuated equilibrium model as a fatal attack on the work of the synthesis. His interest is in an expansion of neo-Darwinism that brings it back into accord with the early pluralism he sees, for example, in the work of Sewall Wright (Gould, 1982a, 1983). If that pluralism has been retained, Gould implies, punctuated equilibrium could be accommodated within an expanded synthesis. But, he argues, the synthesis soon "hardened" into a markedly adaptationist and gradualist canonical version that is now open to severe challenge. Gould professes not to know the causes of this "hardening." He mentions the prestige of solid adaptationist explanations in the field (Gould, 1983, p. 88). These seemed to confirm selectionist experiments conducted in the laboratory, where gene frequencies of different phenotypes could slowly be shifted by controlled environmental variation. We would like to suggest, however, that standing behind this prestige is the deepening influence of empiricist models of science on the synthesists, their successors, and their philosophical defenders. A clean series of inferences linking a mathematically expressible set of laws to laboratory work, and extending laboratory results to the field, was precisely what would be needed to argue that evolutionary theory now stood on a solid, quantifiable, and potentially axiomatic-deductive basis. In particular neo-Darwinists study the evolutionary dynamics of interbreeding populations over successive generations by measuring systematically introduced changes against the stable background of the Hardy-Weinberg Law, which generalizes Mendel's Laws at the population level. In such studies there is often a tacit expectation that selection, rather than processes such as drift, will be the prevailing source of evolutionary change. There is also some expectation that selection will be enforced at a single level, namely, where the organism meets its environment.

The prosecution of this research program, however, has itself turned up, with increasing frequency, results suggesting that selection is not the only source of evolution, or at any rate of gene frequency fixation, and that selection cannot be confined to a single level. The evolution of proteins and nucleic acids seems to manifest a regular pattern of molecular substitutions that does not accord well with selectionist expectations (Kimura, 1979, 1982; King and Jukes, 1969). We have no solid guarantees that similar processes are not at work at higher levels. Moreover, even at the level of study preferred by population geneticists, the now acknowledged pervasiveness of genetic polymorphism accords at least as well, on its face, with nonselectionist sources of gene frequency fixation, such as molecular drift, as it does with selectionist explanations. For although polymorphism has been explained by neo-Darwinists as increasing fitness for changing environments, the rooted expectation of the selectionist paradigm is that genetic variation will be as severely constrained as possible to reflect past and present environmental demands. With respect to selection itself, intriguing arguments have been made that it can occur on, and for the sake of, groups. Other arguments have proposed that selection operates on and for genes and species. (See Brandon and Burian, 1984, for a collection of the most important literature on units and levels of selection.) Finally, the internal integrity and complexity of the genome, which, as several of our authors suggest, is just now coming into view, may force a redistribution of selective pressure to various alternative levels by erecting mechanical or architectural barriers to revision at any one point.

These ideas can be accepted or rejected in various combinations. Gould, for example, has taken up a number of them in his hypothesizing about the causes of punctuated equilibrium patterns of diversification. Chance distributions of genetic variants in peripheral populations— Mayr's "founder effect" (Mayr, 1942)—can trigger sudden and massive reorderings of developmental patterns, provoking speciation (Gould, 1982b). Differential selection can occur on species themselves in a way that is not reducible to selection among their members (Gould, 1982; Stanley, 1979). Finally, patterns of long stability between episodes of radical reordering and speciation can be explained in terms of stable genetic regimes that resist and channel selective pressure (Gould, 1982c). In mentioning these ideas, and the uses to which they continue to be put, it is not our aim to argue on behalf of, or against, any of them. Our concern is to suggest only that there obtains today a great fluidity in evolutionary biology that belies any presumption that selective forces operating at one level are sufficient to enforce explanatory closure on evolutionary problems generally.

These recognitions would perhaps be less troublesome to neo-Darwinists, however, if empiricists had not made much of an analogy between the Hardy-Weinberg Equilibrium and Newton's Second Law.

Both state what will happen in the absence of intervening forces, and so provide a background against which these forces can be measured. This analogy is well conceived. But it has led to suggestions that the Hardy-Weinberg Equilibrium should be no more difficult to apply to biological cases than Newton's Laws to physical cases. This expectation creates a demand for explanatory closure by finding single-level adaptationist arguments that, in their ad hoc quality, can bring adaptationism, and even selectionism, into disrepute (Gould and Lewontin, 1979; Mayr, 1983). Thus while the analogy to Newtonian laws may have increased the reputation of the synthesis for rigor and scope, and so have reinforced hopes that its proponents had found a universal theory of evolutionary processes, it has probably also given a few hostages to fortune. One difficulty is that in biology the move from laboratory to nature is vastly more complex than the analogous move in physics, which has provided its model.

Ernst Mayr points to the fundamental historicity of biological entities as a basic cause of the distinction envisioned here. For Mayr biological lineages are the product of a contingent, well-nigh unique set of historical events. Under the impulsion of causes not readily isolable, they are likely to move off in directions one cannot foretell with confidence. Chance events combine with the entire, unique history of the lineage to make prediction and retrodiction difficult. The fact that this uncertainty can be removed from laboratory populations, we would add, far from putting a solid footing under these vagaries, reflects the fundamental malleability and plasticity of organisms. Thus the move from lab to nature, as Dyke suggests, invites "tricking out" the basic theorems of population genetics with a host of ad hoc devices and ceteris paribus clauses that can, in the end, give scandal to the very laws they are intended to help apply to cases.

Faced with such difficulties, Mayr himself recommends not a renewed search for more powerful and applicable laws, but a retreat from the philosophical demand for them. This demand Mayr views as the outcome of a philosophy of science still preoccupied with an outdated physics and inadequately acquainted with biological research. Most of the putative laws of biology, Mayr says, are really facts. For example, the Hardy-Weinberg Equilibrium is itself the product of the development in evolutionary history of meiotic mechanisms, and thus can hardly count as a basic background of all evolutionary change. (John Beatty has tried valiantly to reconcile this fact with the status of the Hardy-Weinberg Formula as a law of nature by weakening the received notion of such laws rather than, like Mayr, by rejecting it in biology; Beatty, 1980.) Mayr says that the generalizations most useful to biologists trying to establish the facts about some stretch of evolutionary history are embodied in concepts rather than laws. They have an interpretive dimension in the way they are to be applied to particular cases in nature. The interpretive finesse with which the able inquirer brings concepts to bear on individual cases can itself prevent the retailing of ad hoc "just so" stories (Mayr, 1983). Thus Mayr implies here what he has asserted elsewhere (Mayr, 1982): evolutionary biology can and should maintain its deep connection with natural history. But, as Mayr also points out, this view runs counter to the methodological demands of the received philosophy of science. For the interpretive role of conceptual models, on which Mayr and Stent place such great emphasis, is precisely what the progressive testing of hypotheses is supposed gradually to be eliminating from science, as the latter prosecutes its search for ever more comprehensive and basic laws. Faced with a choice between this ideal of science and reality, it is clear, on Mayr's view, that we should reject the physicalist model of science that generates the dilemma.

Whatever the force of this claim, it need not compromise the status of evolutionary biology as a science. It is probably the case that the dialectic between laboratory and world remains a distinctive feature of science among other human practices, and that biological science continues to depend crucially on such a dialectic. But it may also be true that this dialectic takes different forms in different sciences. As both Mayr and Dyke suggest, it is only a contingent feature of this dialectic—the outcome of cultural circumstances surrounding the development of physics as the paradigm of science generally—that we have come to regard nature as one vast laboratory, whose parameters will eventually come under our control by our grasp of fundamental laws from which all else can be calculated. The original Darwinian paradigm honored this view by regarding all natural populations as at or tending toward Malthusian competitive limits. This commitment is maintained by single-level selectionist versions of the synthesis. However, according to Dyke, the discovery of a degree of complexity that bursts the bonds of this paradign—a theme broached by many contributors to this book—calls for a shift to treating the central investigatory generalizations of biology as "ideal types." These can impose only hypothetical "closure conditions" on an object of inquiry, rather than the categorical closure conditions of a fully reduced system of law-determined nature. In insisting that the same holds for physics, Dyke thus argues that devices adopted by post-Weberian social scientists as second-best substitutes for the laws they had hoped for, but failed to find, are in fact intellectual tools of the first rank. The utility and power of these tools in biological applications show that they need never have been invidiously contrasted with strict laws. Gunther Stent's reflections on the role of hermeneutics in developmental biology constitute the most radical testimony to this view contained in this book.

The complexity attendant upon the historicity of biological systems can only be intensified by recognition that such systems are fundamentally and nontrivially hierarchical in nature. For this means that events occurring at one level of a system cannot be deductively or predictively inferred from processes known to be occurring at other levels. In addition to ruling out the possibility of full reduction, this fact might appear to make it impossible to achieve any solid measurement. Indeed, the problem of measurement has for some time been the preoccupation of the best minds in evolutionary biology (Levins, 1966, 1968; Lewontin, 1974; Wimsatt, 1980). Yet ironically it is precisely the fact of the hierarchical nature of living systems that allows a solution to the problem of measurement in the face of complexity.

Conditions for getting robust results from the application of mathematical models to biological cases depend on the possibility of "sealing off" events (in a sense developed by H. Simon, 1962, 1973) at one hierarchical level from influences coming from another. This allows explanatory closure to occur in particular cases at a given level of investigation. Since it is precisely the hierarchical nature of biological systems that allows this sealing off to occur, hierarchical structure, rather than being a feature that stands in the way of getting robust results, actually makes such results possible. This opportunity can be seized, however, only when reductionistic ideals are rejected, for what must be foregone in order to accept such measurements is any expectation that events at one level will be shown to be a simple function of events at another. Following Levins and others, Marjorie Grene has spoken of choices that must be made among incompatible scientific values. Here is precisely such a choice.

At this point, then, it will be useful to reflect briefly on the properties of hierarchical systems. In a recent study, Allen and Starr describe a hierarchy as a system of structural or behavioral interconnections, wherein higher levels constrain lower ones (Allen and Starr, 1982). The lower levels may be component parts of the higher (as cells to tissues), in which case they speak of a "nested hierarchy," or may be independently individuated entities nonetheless constrained in some important way by higher levels of organization. Allen and Starr cite the relation between resources and consumers in an ecological community as an instance of such a "non-nested" hierarchy. One important feature of hierarchies, they maintain (following Simon and Pattee), is that separate levels tend to possess distinctive cycle times for the recurrence of events. The periods of such cycles are often several orders of magnitude apart. Cycle times thus give us a reliable guide to separating out levels of a hierarchy, measured in terms of direction and rate of information flow between the levels.

According to Allen and Starr, information in a hierarchical system typically flows asymmetrically from higher to lower. That is why the higher levels of a system provide a relatively stable environment within which endogenous behavior at lower levels occurs without destabilizing higher levels. To these reflections we would add, however, that lower levels still exert important constraints on higher ones. Indeed, past neglect of the structural autonomy of higher levels, which allows for considerable independent activity at lower levels, derives in good part from the power of lower levels to constrain higher ones in simple physical systems. Rejection of reductionist programs in favor of a hierarchical view of most of the interesting systems of nature should not lead us to neglect the fact that while higher systems cannot be reduced to lower, they are nonetheless importantly tied down by and to them.

One of the most crucial questions about hierarchical systems arises from the distinction between nested and non-nested hierarchies. How are we to distinguish individuals in relation to groups from parts in relation to wholes? Certainly, as Allen and Starr recognize, our own perceptual-cognitive history provides us with a familiar and powerful way of doing this. However, as David Hull has pointed out in a seminal paper (Hull, 1980), to get a comprehensive evolutionary theory we may have to abandon our intuitions in this matter. If selection occurs, for example, on species, we may have to regard species as individuals rather than as groups, and organisms as parts of species. If we do not, the logic of selectionist arguments, which is focused on differential fitness among individuals, may on Hull's view be compromised.

Dyke proposes that in this matter we allow the needs of inquiry and the prospects of a productive research program to be our chief guides. For the perceptual apparatus we have evolved for individuating and counting objects is on no very different epistemological footing from the theories we have or might develop by sustained theoretical discussion. Each is a system of identification and orientation competent in some areas and incompetent in others. Our perceptual apparatus, in the very nature of the case, is almost totally insensitive to very high and very low cycle times, and so is by itself incompetent to inform us whether the entities involved in such cycles are individuals or parts. This is both a liberating and an unsettling insight. It suggests, in fact, an analogy between the present condition of evolutionary biology and the situation of physics earlier in this century, as Hull has seen (Hull, 1980). The change from a Newtonian world view came about when physicists began investigating very small and very large objects. The crisis was not resolved until ontological and epistemological revisions had resulted in an alteration of our conceptions of space, time, matter, continuity, and causality. Neo-Darwinism appears to be undergoing a similar crisis at the present time, a crisis that will not be resolved without equally significant conceptual revisions.

The more supple methodological and ontological perspective we have just been reviewing has been suggested by the irreducibly hierarchical nature of the biosphere. We now see that a hierarchical view of the biological world not only reinforces the claim that explanatory closure cannot be enforced at a single level, but invites us to consider positively the possibility that selection may occur at and across any number of biological levels. For where closure at a single level fails, nonselectionist processes are free to operate and to provide a platform on which selection can go to work on one or more new hierarchical levels, whether the objects at those levels are regarded as groups or as individuals. A hierarchical expansion of neo-Darwinian thinking incorporating some of the ideas we have reviewed has been proposed by Arnold and Fristrup (1982). The possibility that neo-Darwinism can be freed from an historically contingent connection to reductionistic programs in the philosophy of science depends on the success of such work.

Contemporary Reductionist Programs

The fact is, however, that while this program for an expanded neo-Darwinism is inconsistent with a reductionistic philosophy of science, many scientists are still attracted by reductionism. Their preference can be defended in terms of at least one scientific value that their opponents will usually accept—the long record of success in programs that have at least driven toward reductionistic explanations, even if they have not completely found them. Methodological reductionism is often a valuable, sometimes an indispensable tool in getting us to see how things really work. The only difficulty is the entanglement in ontological reductionism that so often attends the drive toward underlying mechanisms. This entanglement is a theme broached by several essays in this book, especially those of Brandon and Dyke. A word about this topic is thus in order.

An intriguing case of a radically reductionist reconstruction of evolutionary biology has been presented by Richard Dawkins, who reaffirms the connection between the Darwinian tradition and reductionism by giving up one linchpin of that tradition. Instead of arguing that it is the organism that primarily benefits from genetic changes that increase fitness, Dawkins has suggested that successful interactions between the organism and its environment benefit self-replicating stretches of genetic material—genes in an extended sense—and benefit them in the rather strong sense that organisms and their organs are there in order to advance the self-replication of the genes themselves. As Dawkins has colorfully put it in *The Selfish Gene* (1976), organisms are the gene's "survival machines." Generalized by way of David Hull's perspicuous terms (Hull, 1980), the point is that *replicators* and not *interactors* benefit from adaptedness. The problem of measurement can then be transferred exclusively to the genetic level in a way that accords with, indeed radicalizes, the call of G. C. Williams for an evolutionary theory that is as parsimonious and reductive as possible (Williams, 1966).

Difficulties abound with this program (Sober and Lewontin, 1982; Wimsatt, 1980; both are reprinted in Brandon and Burian, 1984). Not the least of these is the failure of Dawkins to give proper ontological value to objects at transgenetic levels, a topic explored by Dyke in a related connection. In Dawkins's work we see theory reduction and epistemic reduction anchoring themselves in ontological reduction. Robert Brandon is particularly concerned, however, to show that Dawkins's approach runs roughshod over another important aspect of evolutionary theory: the meaning and role of the teleological or purposive ascriptions that are used to make adaptationist explanations.

In most empiricist philosophies of science, teleological ascriptions are the very paradigm of the arbitrary imposition of a priori conceptual schemata on the raw data of experience. Even Kant agreed that such conceptions could never form a matrix for genuine knowledge, however inevitable or useful they were in our thinking. Thus, if apparently purposive phenomena are not to be thought of in terms of intentional purposes or vitalistic mysticism, they must be shown, according to empiricists, to be complex cases of the same explanatory patterns obtaining in more basic sciences (Nagel, 1961, 1977). However, Brandon argues here and elsewhere (Brandon, 1981) that this demand is an unnecessary assault on our intuition that the biosphere does in fact contain systems whose genuine purposiveness expresses the fact that they have come about in a different way from physical systems. The key to Brandon's argument is his insistence that when we ascribe a function or purpose to some organic feature, we make not only a descriptive remark, but also, properly speaking, an explanatory one about it. We state why the feature has come to exist and to work the way it does; and we do so, in adaptationist explanations, in terms of the effects of past instances of the feature as causes of new instances of it by way of differential fitness. The acceptability of these explanations implies important differences between physical and biological evolution. For the systems that are the subject of the former cannot be the subject of questions that take answers in the form of functional and purposive ascriptions, while those of the latter can. In arguing against Dawkins, Brandon shows that explanations of this sort must take interactors rather than replicators to be their subjects at all hierarchical levels. Thus Dawkins's teleology of replicators fails to conform to the conditions under which teleological explanations in biology make sense, and so wavers uncertainly between mechanism and mysticism. Here a program of genetic reductionism seems, then, to threaten the most powerful explanatory achievement of the Darwinian tradition, rather than to render that achievement more secure.

A second recent reductionist program is that of such sociobiologists as Hamilton, Trivers, Barash, and E. O. Wilson. Sociobiology of this stripe makes extensive use of the notion of inclusive fitness, according to which adaptations may arise that benefit not only the individual organism but the kin of that organism to the extent that they share its genes. This theory depends on the applicability of game-theoretical mathematics to evolutionary biology. It is, as Dyke notes, the latest example of a long Darwinian tradition of borrowing from mathematical techniques first applied in economics. This idea has been successful in producing plausible explanations of many phenomena that appear to be more altruistic than a model of individual organic competition will permit. In particular, Wilson has been so successful in applying this model to the evolution of social insects that he has been emboldened to generalize his results to human societies (Wilson, 1975). However, as a condition of the applicability of game-theoretical mathematics to biology, it can easily appear that genes must at all times be seen as attempting to maximize their own persistence. This is in fact the source of Dawkins's genetic teleology. But this conception seems to carry with it renewed commitment to the grim Malthusian competition of early Darwinism, and so a retreat from the success that neo-Darwinism has had in moving away from it. For neo-Darwinism, with its stress on relative or differential fitness between lineages, does not strongly assume that lineages are always pursuing a maximization strategy at Malthusian limits. Indeed, it is precisely because of the relative openness implicit in neo-Darwinian thinking that the synthesis can be expanded, as we have seen, to accommodate selection at a multitude of levels. If closure were enforced at one level, as it was at the organismic level in early Darwinism and is at the genetic level in recent reductionistic programs, this expansion would be impossible.

Marshall Sahlins has drawn attention to the ideological implications of this crypto-Malthusianism (Sahlins, 1976). He believes, however, that the way to resist Wilson's argument is to draw a line between the natural and the cultural. Notwithstanding how things may be with the bees and the ants, human culture "interrupts" nature, and so interrupts the chain of inferences that Wilson would like to draw. The point of Dyke's essay, however, is to suggest that the problem does not lie with the possibility of a human sociobiology, or with the use of game theory in biology, but with reductionism. It is only against the vision of a fully reduced nature, all of whose dimensions are controlled by events at a lowest level, that we need to preserve our intuitions about the uniqueness of human culture by withdrawing it from nature. A fully articulated, nonreductionistic, hierarchically structured naturalism might well be comfortable with the general notion of sociobiology, and will find game theory helpful without importing Malthusian closure conditions.

For a variety of reasons, then, it would seem that progress in evolutionary theory, and its relevance to the human sciences, can be secured only if philosophical pressures leading to substantive reductionism have been alleviated. In the following section we shall suggest a perspective on science that we believe is amenable to this purpose.

Sketches for a New Philosophy of Science

The challenges to orthodox neo-Darwinism explored above have come into prominence at a time when the philosophical atmosphere differs *toto caelo* from that which prevailed when the Modern Synthesis established itself. Just as the reception of neo-Darwinism was colored by prevailing empiricist currents of thought, so too recent inadequacies in the synthesis have sometimes been seen as carrying with them the threat of a paradigm crisis in evolutionary biology. They have thus been peripherally affected by the widely known work of Thomas Kuhn.

Punctuated equilibrium, for example, has been presented by Gould himself as a challenge to the gradualist ideology that stands in the background of both the old and the new Darwinism. Though an approach like this offers an opportunity for much insight into the deeper conceptual assumptions behind even the most precise and powerful theories—insights to which empiricists are often insensitive—it nonetheless imports its own difficulties. Will a new paradigm do nothing but establish a new ideology in place of the old? Does sensitivity to the sociological dimensions of scientific theories lead to rejection of science's claims to tell the truth about the world? Are scientific explanations ideologically contaminated? If so, then a new philosophy of science can make resuscitation of the old empiricism seem attractive, even if it brings in train a reductionism that must remain uncomfortable with the historical, purposive, and hierarchical nature of biological systems. To avoid these consequences we will do well to try to discover whether a third alternative might be available.

A generation ago Thomas Kuhn (1962) published his well-known argument that scientific knowledge does not and could not accumulate in the way empiricists had thought. He thus initiated a process of fact-finding and soul-searching in the philosophy of science that has not yet run its course. Kuhn's study of the revolutionary episodes in the history of science to which empiricists looked back with pride suggested to him that change of belief in scientific communities was, at critical moments, not unlike change of belief in religious, political, and aesthetic-interpretive communities. In all cases, antecedent acceptance of an interpretive framework allows one meaningfully to apprehend and assess data. Because such frameworks, or paradigms, are often powerful enough to suppress potentially disconfirming anomalies, at least for a time, acceptance of this thesis has suggested to some a scepticism about scientific knowledge. What such interpretations have in common, how-

ever, is a tendency to assume, with positivists, that nonscientific discursive practices are fundamentally noncognitive in nature, and so
nonrational and nonself-correcting. For the revelation that change of
belief in science involves interpretive practices like those common in
other fields leads to scepticism only when science is thereby seen as
infected by an arbitrariness and subjectivity from which it was previously
thought to escape.

It is worth noting that Kuhn has sought to dissociate himself from
these implications (1970a, b). The quickest way to do this is to take
some initial distance from the assumption that human discursive practices, unless interrupted and replaced by strict, methodologically controlled inquiry, are presumptively noncognitive. The scientific world
view brought with it out of the crisis of early modern thought just such
a jaundiced assumption. To repudiate that view is thus to cut the ground
from under the old idea that only science offers us security in a world
otherwise full of snares and delusions. If we take just enough confidence
in the cognitive worth of human discourse generally, we may in fact
be able to renew the effort to see what is especially (rather than uniquely)
cognitively powerful about science.

The new or post-Kuhnian philosophers with whose efforts Marjorie
Grene associates herself share this confidence, although agreement
among them on the consequences of this initial shift in perspective is
still very piecemeal and inchoate. By commenting on points made by
Grene and several other philosophers who have contributed to this
book, we can begin to see how one might arrive at a view of science
that retreats from reductionistic ideals without sacrificing the possibility
of scientific truth and progress. Such a view cannot help being relevant
to current problems in evolutionary biology.

According to Grene, the most basic component of a more adequate
philosophy of science is a deeper view of perception, and of its relation
to knowing, than empiricists have given us. Grene agrees with the
empiricist tradition in holding that knowing is rooted in perceiving.
But she notes that empiricism favors a view of perception in which
discrete perceptual units, which impinge on a passive percipient organism, are ordered in terms of highly abstract conceptual structures.
Given this view, someone might say: If our conceptual structures are
not as formally empty as empiricists require (and they are not), then
our conceptual inheritance will overwhelm our weak perceptual foothold
on the world and lock us into the echoes of our own projective world
views. What is required to undercut this impasse, then, is a richer view

of perception, in which perception is seen as an active process, an achievement, and already cognitive, rather than as a passive foundation on which something else, either true or illusory, can be built.

Many new philosophers of science would agree that an alternative to empiricism and relativism should be built on a better account of perception and its role in science (Brown, 1977). Grene's exploration of this idea is especially helpful, since it is rooted in a view of man as a biological being, much of whose cognitive history is itself a biological accomplishment. Human beings, no less than other animals, interact with their environment by way of relatively holistic perceptual schemata—schemata that also play an important role in Gunther Stent's essay—in terms of which they actively orient themselves within their world. On this view, it is precisely our own natural history that underwrites the cognitive worth of human praxis in general and of scientific praxis in particular. For it is our own evolution that has created these schemata, through which we engage in an active, exploratory dialectic (or, as Stent would say, hermeneutic) with the world, matching the unknown against our perceptual models, correcting when experience does not fit our anticipations, and in the long run incorporating new, more reliable, more comprehensive structures into the models themselves. If we are careful enough, we may see our historical learning as a continuation of learning patterns already established in and through our perceptual experience. Just as our perceptual experience is already incipiently conceptual, so too our shifting and changing conceptual schemes are the products of a deep dialectic with perception and with previous conception.

Although great care must be taken in filling out this picture, it already contains several significant implications for the philosophy of science. It implies, as Grene is quick to point out, that we need not generally censor our natural presumption that our theories can, and in some measure do, accurately comprehend the world. For while Grene's "comprehensive realism" does not resolve all the issues associated with the problem of "scientific realism," it does remove an important barrier to realism by letting us see that we are more deeply, and hence more advantageously, placed within nature by our perceptual and intellectual abilities than either empiricists or their opponents have led us to believe. The breakdown of the adequacy of our theories in particular cases should therefore not lead us to suspect our ability to know the real world. On an abstract view of perception and an impositional view of conception, on the other hand, such general doubts might well be natural on the heels of specific ones.

This picture gives us a new perspective on scientific change, progress, and cumulation. If our embeddedness in history is structurally similar to and continuous with our embeddedness in nature, we may come to see that even the most sharply competing paradigms arise out of a tacitly common background of understanding. That is why, in the long light of historical reconstruction, we can see much continuity where contending theorists may have seen only disagreement, and a gradual accumulation of changes that culminate in real ruptures. These ruptures preserve many of the changes that provoked them, while destroying conceptual forms that can no longer contain them. Scientific change can accordingly be rational, even where it isn't antecedently calculable (Toulmin, 1972).

Once this point has been grasped, it is to be expected that theoretical terms first deployed in one conceptual matrix can be taken up, through a process of intensive interchange between members of one scientific community and another, into a new one, and ultimately absorbed into the practice of detailed scientific inquiry, where they have a life of their own (Hacking, 1983). Thus it should not surprise us that Richard Burian, in his study of the interchanges between Bateson, who did not think that genes were physical entities, and Morgan's research group, who learned that they were, should fail to turn up the mutual misunderstanding and referential failure predicted by relativists. What prevents recognition of this communicative success, Burian suggests, is largely the demand for a guaranteed method of achieving continuity. According to Burian, this demand is shared by empiricists and their sceptical opponents. When we see that risks to continuity in science are no less than those that attend serious discussion within and among other sorts of discursive communities, we can come to see how, and how often, continuity is in fact achieved. We need not, then, draw radical conclusions from the breakdowns of understanding and agreement within and among scientific communities that do sometimes occur. Where institutions the very point of which is to press forward have been set up within a culture, we may expect that renewed and advanced understanding will emerge.

From these confident presumptions—the very opposite of the sceptical doubts that so troubled Hume—we can already draw one important consequence about reductionistic ideals. If we can presume a comprehensive realism that is not unduly troubled by global doubt, we need not insist that scientific praxis must, on the one side, compulsively drive toward a pre-interpretive, preconceptual bedrock of sensory data or observational reports on which its claims to truth must be founded; nor, on the other, must we construe its progressiveness as a function of a dutiful push toward a telos in which method will have purified data of all interpretive structures. By the very intensity of its discursive involvement, scientific inquiry pushes our total perceptual and conceptual inheritance toward greater integrity. Seen in this light, philosophical pressures toward reductionism, springing from programmatic anxieties about knowledge rather than from real questions that arise in the course of scientific inquiry, can be discounted and largely ignored.

We should not, therefore, try to envision the life of investigative communities as centered on attempts to find cognitive guarantees. We are not likely to uncover much about the actual discursive practices of such communities—how they conduct their discussions, how their discussions relate to those of other such communities, how the discussions of the scientific community as a whole are related to other discursive communities within the larger society—if we color our inquiry with these worries, and think of scientists as bearing the epistemological weight of the world on their shoulders. If, on the contrary, we approach such questions with the more relaxed attitude we are recommending, we are likely to turn up additional points about the inappropriateness of reductionistic ideals.

We may note, for instance, what new philosophers of science have held and Grene has reemphasized: that a marked characteristic of scientific communities is that they are collectively bent on the solution of real problems (Laudan, 1977; Hacking, 1983) rather than preoccupied with larger epistemological worries or programs. The relevant conception of problem solving differs, however, from empiricist approaches to this topic in holding forthrightly that more is involved in successful scientific problem solving than merely deciding which of several competing frameworks better predicts a pattern of data. Increased conceptual coherence for its own sake is also a value that is pursued. At root, this is because conceptual and empirical aspects of experience are not fully separable, having evolved together in both nature and culture. But it is also because conceptual coherence in science, no less than in other areas of human practice, is an intrinsic good, even where it is initially unaccompanied (so Laudan argues) by greater empirical power than its competitors. To admit this permits one to see dimensions of scientific continuity and progress that are inaccessible when only empirical con-

siderations are thought to be relevant. In particular, it offers an opportunity to see continuity in science as achieved by conceptual revision
rather than interrupted by it.

It follows from the importance of conceptual discussion in science
that the argumentative practices of a scientific community are not a
replacement for, but rather an intensification of, sound argumentative
practice in all fields open to rational discourse. More is required in
scientific problem solving than scrupulous fidelity to received canons
of experimental inquiry and reporting. These practices will be instan-
tiations and intensifications of wider standards of rationality applicable
to all kinds of inquiry. These standards include sensitivity to tacit assumptions, subtlety in judgment about particulars, and quickness to
discern the implications far and near of a concept or a principle.

Brandon's use of "erotetic logic"—the logic of questions and answers—to elicit the precise nature and role of teleological discourse in
evolutionary biology exemplifies recognition of these dimensions of
scientific discourse. It implies, in the first place, that the formal languages
of science are not alternatives to the semantical opportunities and demands of natural language within and through which human intelli-
gence has evolved, but render these more precise. From this perspective,
Brandon is able to elicit important distinctions between physical and
biological evolution, and to ratify our meaningful talk about purposiveness. His work also suggests why and how scientific inquiry can
make some room for interpretive methods of inquiry still too exclusively
identified with nonscientific discursive communities.

The increased conceptual coherence that attends progressive problem-
solving within a scientific community often centers on the adoption or
presumption of fundamental philosophical assumptions of a kind that
bridge scientific and nonscientific thought (Laudan, 1977, pp. 101ff).
Bringing such assumptions within the scope of scientific theorizing
sometimes provides an occasion for the revision of received religious,
political, and cultural beliefs. That is one of the most exciting things
about science. But by the same token we are not entitled to initially
write off the cognitive import of our nonscientific beliefs, for we understand these matters too, and indeed the overall coherence of our lived
experience, in terms of an inextricable integration of experiential and
conceptual considerations. We must therefore admit that the acceptability or nonacceptability of our scientific commitments is affected by
their overall integrability with other aspects of our understanding. Such
a conception is fundamentally at odds with the assumptions on which

the received philosophy of science has been based. For it challenges the axiom that other forms of human praxis have no cognitive standing that is not first extended to them by science. It holds, on the contrary, that there can and should be reciprocal interchange among all the various dimensions of human understanding.

This is not to deny that scientific praxis should limit its openness to other aspects of culture during the working out of a promising and productive research program. It is to say rather that the appropriate degree of openness and closure is a matter of judgment. It is no less dogmatic for a community of scientists to categorically delegitimize the discourse of other communities than for such other communities to force scientists to open themselves at every moment to ideological winds that never cease blowing. Dyke thinks these points are of particular relevance to evolutionary biologists, since the roots of their discipline lie in an intricate dialectic with natural theology, and with a culture to which natural theology was existentially important.

In the light of these reflections we may once again approach the topic of reductionism. It is true that discussions between scientific communities sometimes coalesce to such a degree that we are tempted to speak of the reduction of a theory achieved by one community to that achieved by another. However, as Darden and Maull (1977) have shown, even in the most obvious cases two fields are fused by way of "interfield theories" that in the process change both the "reducing" and the "reduced" science. It is often the case that the relevance of theories developed in one community to problems puzzled over by another is, however, looser than that. In general, we may imagine a flowing continuum of relevance and irrelevance obtaining among various communities. It is a dogma of empiricism, however, that progress in human knowing is indexed, indeed held hostage to, success in shifting discussions at the looser end of this scale of relevance among discursive communities toward the tighter end. Only the fact of this tendency, so the argument goes, gives us any assurance that there is such a thing as human knowledge at all, in contrast to the shifting tides of fantasy and fallacy that characterize most human affairs. Once again, then, we see that the programmatic demand for a reductionistic unity-of-science is propelled not just by highly general dogmas, but by a pervasive and diffuse tendency toward scepticism imposed upon the inner working of scientific communities. This program is far from obviously reflective of the real nature of the interactions among and within such communities of inquiry.

It may also fail to reflect the inherent relations among the subject matters studied by each. Though our present division of labor in the life sciences can hardly reflect very well the ultimate nature of things, the fact that different scientific communities preferentially focus on different dimensions of the real might serve us well in a biosphere constituted of partially decoupled hierarchical systems. The irreducibility of the discourse of one such community to that of another, far from being a residual surd that must perhaps be grudgingly accepted (Popper, 1974) would then help rather than hinder our attempts to penetrate further into the complex architecture of the world. Reductionistic solicitations born of epistemological worries might, by the same token, have the opposite effect.

The accumulation of knowledge under these conditions is best served when a high degree of readiness exists among scientific communities to discern the relevance of work done in fields other than their own. At present, for example, the mutual relevance between molecular biology and population genetical evolutionary theory is quite high. With their vastly different histories and hopes, we may expect as much heat as light to be generated by this exchange. But out of this interaction we may also trust that much new understanding will emerge. The only thing we can be reasonably sure of is that the convictions of one community will not have been reduced to those of another. This remains true even if each community maintains a preference for the most parsimonious methodological strategies appropriate to its subject matter. In the final section of this chapter we will explore further the emerging exchange between molecular biologists and population geneticists, and in the process attempt to suggest something more about the nature of scientific change and continuity.

The Limits of Selection and the Future of Darwinism

In this section we argue that the philosophical perspective urged above can be of help in assessing the significance of claims emerging recently from molecular genetics that nonselectionist mechanisms play an important, indeed a basic role in evolution. However it is articulated, this possibility challenges the selectionist focus of the Darwinian tradition, even when the latter admits some role for nonselectionist processes and for selection at and across a number of levels. It thus raises questions about the correctness of evolutionary theory that are perhaps tougher than those posed by the effort to expand the selective paradigm itself.

Accordingly, it poses more sharply the question whether there can be conceptual continuity between the Darwinian tradition and whatever view of evolutionary processes emerges from discussions now under way.

We may begin with a brief review of work on evolution carried out in recent years from the perspective of molecular biology. The possibility, recognized by neo-Darwinians themselves, that gene frequencies could be fixed by random processes as well as by selection, has been reaffirmed and extended by molecular evolutionists. Analysis of numerous protein structures and enzyme electrophoretic patterns suggests that a large number of mutations may be selectively neutral at the molecular and physiological levels. This could explain the extensive polymorphisms observed at many gene loci at least as well as balancing selection does. King and Jukes (1969) concluded on these grounds inter alia that natural selection was not nearly as important as had been assumed by neo-Darwinists; indeed Kimura (1982) has extended their analysis to claim that the vast majority of evolutionary events occurs by such selectively neutral changes.

Moreover, the neutral hypothesis predicts that since protein mutations are fixed stochastically, the rate of mutation should be constant over time. A. C. Wilson (Wilson et al., 1977) has shown that there is in fact no simple coupling between molecular and morphological evolution. Much lower-level alteration is irrelevant to morphologically discernible change, while the constant rate of change at that lower level suggests that it may be immune to selection.

More recently this relatively static conception of nonselective evolutionary change has been overshadowed by intimations of more dynamic, yet still nonselectionist, evolutionary mechanisms. Molecular biologists such as Dover have used terms like *molecular drive* (Dover, 1982) to name, though hardly to analyze, this possibility. We now know, for example, not only a great deal about the repressor-activator mechanisms that regulate gene expression in prokaryotes, but we know too that prokaryotes can, inter alia, exploit transposable genetic elements, and exchange information via plasmids. More astonishing has been the discovery that eukaryotic genes are not just stretches of passive DNA, in which fortunate mistakes occasionally happen. Rather, many eukaryotic genes have a complex structure, often with many copies in a multigene family (Campbell, 1982, 1983). Indeed, the eukaryotic chromosome should be viewed as an organelle in its own right. Moreover, virtually all eukaryotic structural genes contain intervening noncoding sequences, introns, which recent studies suggest may have enzymic activity and which have to be processed out to make mRNA. These may play an important evolutionary role in shuffling the protein domains coded for by the exons. Further, there are a number of proteins that interact directly with the DNA, plus a complex enzymatic machinery that can manipulate the genetic message, altering its direction or position, regulation, and function. The virtual zoo of enzymes that manipulates DNA is in fact responsive to hormonal signals and can reflect environmental conditions.

Thus the genes most likely to be important in evolution are, in Campbell's terminology, "profane," rather than the "sacred" genes postulated by the neo-Darwinian interpretation of Mendelism. That is, the gene is not approximately invariant in structure but can readily be manipulated by the cellular machinery. Some genes even "cheat" to beat Mendel's rules. Thus the genome can be seen as part of a larger information processing and modeling system (Guilleman, 1978; Campbell, 1982, 1983; Campbell and Zimmerman, 1983). In his contribution to this book, Campbell goes on to suggest how this view allows directed adaptation and, in the most advanced evolutionary forms, adaptation to a projected future environment. This view virtually reverses the previously assumed relationship between adaptedness and evolution in which, according to orthodox neo-Darwinians, the latter is always a product of the former.

This sophisticated neo-Lamarckianism might provide an alternative mechanism for the rapid, "saltational" evolution postulated by Gould, as well as for the prolonged stasis that is equally esssential to his argument (Gould, 1982). As always, however, there is a price to be paid. To accept this view, we must abandon or mitigate the Central Dogma of molecular biology, according to which information is construed as passing exclusively from DNA to protein, and not the reverse. The Central Dogma would thus reveal itself in hindsight as what Dyke has called a "sealing off" assumption, the utility of which collapses with the success of the very research program it has made possible.

In a review of ideas currently emerging from molecular genetics (Lewontin, 1982b), including those of Campbell, Richard Lewontin does not fail to note irony in the fact that molecular biology, a discipline at first aggressively committed to fully reducing biology to chemistry and physics, is now generating challenges to received evolutionary thought that are the very opposite of reductionistic, and that suggest how far neo-Darwinism itself has been limited by its own misplaced regard for reductionistic ideals, even while resisting its own reduction to physics.

On Campbell's view, none of the ideas just reviewed can be coherently deployed without expanding the role of teleology beyond the logic of consequence-selection so ably explicated by Brandon. He believes that a more ambitious purposiveness is required, in which consequence-selection itself favors capacities for further evolution that are fixed antecedently to specific environmentally useful adaptations and thereafter live a life of their own. To envision such processes is to transcend the logic of teleological arguments that has been developed thus far by the Darwinian tradition. His view seems in fact to recall older, anticipatory conceptions of teleology that were explicitly rejected by Darwin at the outset.

Meanwhile, Stuart Kauffman's approach to genomic autonomy travels another road not taken by Darwinism. Kauffman is convinced that if we explore evolution only by remaining at the environmental surface where chance and purposiveness combine to produce historical uniqueness, we shall have to abandon hope for a rigorous, predictive evolutionary science. Unlike Mayr and Stent, Kauffman is loath to compromise these scientific values. Thus he is pleased to find internal factors deep within the structure of the genome that can sustain robust predictions on the basis of statistical laws. When connectivity within the regulatory system of the genome is rich enough, selection will be constrained by generic features around which an evolutionary ensemble varies, and toward which it tends to return. Here, then, is a stable background of ahistorical universality against which the historicity of evolutionary paths can be measured, and in terms of which the maintenance of stable characteristics can be explained. Explorations of evolution from the point of view of nonequilibrium thermodynamics are analogously motivated (Wiley and Brooks, 1982; Brooks and Wiley, 1984; O'Grady, 1984). Endogenously generated informational crises in ontogenetic programs are said to work with selection, or even independently of it, to produce speciation. Perhaps biology can move closer to the law-governed paradigms of physics if physicists reflect upon the startling properties of systems far from equilibrium (Prigogine, 1980; Prigogine and Stengers, 1984). The appeal of teleology might well diminish in proportion to the success of this research program (O'Grady, 1984).

The reflections in this section have a bearing on the important issue of the link between ontogeny and phylogeny, the solution to which will highly recommend any evolutionary theory that gives a key to it.

Kauffman's approach to the problems of developmental biology contrasts with that of Stent. Both Kauffman and Stent reject the old notion that ontogeny is programmed into DNA by selection. It is not on such reductionistic terrain that molecular biology, evolutionary theory, and developmental biology are likely to meet. But, according to Stent, this fact invites hermeneutical inquiry into processes that are fundamentally historical rather than law-governed, while Kauffman thinks that developmental sequences reflect statistical properties of basic evolutionary ensembles that may indeed be comprehended by laws.

For some time Mayr has argued that a rejection of "essentialism" in favor of an emphasis on individual differences is the intellectual source of the Darwinian revolution itself (Mayr, 1976, 1982). The individuality and historicity of organisms calls, in this view, for hermeneutic rather than nomological modes of inquiry. Perhaps, however, resistance to essentialism in Mayr's sense has also inclined Darwinians to keep their distance from law-like patterns of universality such as those to which Kauffman wishes to draw our attention. For universals have traditionally been identified with essences. Kauffman's conceptual recommendations, therefore, like Campbell's, urge a reappropriation of conceptual tools that the Darwinian tradition has generally forsworn.

We may see from these reflections that conceptual and methodological strains already noticed within contemporary neo-Darwinism are to be found in an intensified form where the neo-Darwinian inheritance meets advanced molecular biology. However, amid attendant disputes about how best to proceed, and what conceptual structures to employ, important facts are beginning to emerge. Kauffman and Campbell would agree that lineages evolve by developing capacities best represented as heuristic rules inscribed in the regulatory system of the genome.

These considerations imply, however, that the relations between selectionist and nonselectionist processes in evolution are complex and various. We have already seen that an expanded neo-Darwinism, which distributes selection across a number of hierarchical levels, allows a role for nonselectionist processes. We now see that selection may be deeply constrained by nonselectionist universals or, alternatively, may be guided by anticipatory capabilities. In either case, the received conceptual apparatus of the Darwinian tradition is strained, and progress invites conceptual revision and innovation no less than renewed empirical efforts. Perhaps the notion of selection itself can be reworked to accommodate the facts to which Campbell and Kauffman point. Perhaps its role must be pruned back. Whatever the outcome, we are entitled to ask at this point what the implications might be for the continuity of evolutionary theory.

To explore this question we might recall certain features of Darwin's own paradigm, seen within the cultural space in which it was proposed and defended. Darwin's theory was an explicit extension of the Newtonian paradigm to the biosphere, inspired by and utilizing its earlier successes in economics and geology. Basic to this paradigm is a conception of nature as a closed system in which changes in one variable strictly entail corresponding and measurable changes in another. To the pressures that an ever-changing geological environment would bring to bear on organisms, Darwin added the greater pressure of scarce resources on the inherent dynamics of population growth, as postulated by Malthus. He went on to argue that these combined pressures would be fatal to the future of organic lineages unless the characteristics of better adapted variants were inherited and amplified to maintain adaptedness by way of species transformation. Darwin's world was thus a pressure-cooker world governed by a closed system of natural laws. Were it not, it would, he felt, still allow natural theologians to assert that no set of natural forces would ever be sufficient to account for the intricate functional organization of plants and animals, and their adaptedness to their environment. In a cultural setting whose general terms of discourse appeared to make such assertions not only intelligible but plausible, the point of Darwin's work was to show that a viable alternative could be conceived.

Recent research has suggested (Ospovat, 1981) that in the decades prior to 1859 Darwin may have rested in the orthodox Newtonian view that the system of natural laws, in biology no less than in physics, formed a set of secondary causes to carry out the Divine Will, without being guided by that will in matters of detail. The principle of natural selection could thus execute God's aim to produce, by operation of natural law itself, an evolutionary creature who could discern these very processes with his very well-adapted but fully natural mind, and so engage in a free man's worship (Ospovat, 1981). But if this progression was not to be a rigged game, the meeting between genuinely chance variation and environmental utility must, in the end, freely move in whatever directions circumstances took it, rather than toward some preordained outcome. Recognition of this fact would have progressively led Darwin to his mature view of a world that may well have grandeur but has no overall aim. Variation, arising independently of adaptive utility, would be taken up at a deterministic Malthusian surface, where organism and environment meet, to produce a constantly changing lineage whose future course would be affected by its entire earlier history, but by no telos. To further blunt the possibility of a directed teleology, however, Darwin would have been inclined to inscribe his theory, however secretively, within the bleak parameters of a reductionistic materialist world view (Gould, 1977). For only such a picture could, in Darwin's cultural circumstances, provide a guarantee against the return of transcendent premises or conclusions. Despite the fact that the deterministic and reductionistic implications of that world view did not fully accord with his stress on a chance relationship between the levels of variation and adaptation, Darwin's purposes would constantly pull him in this direction.

It is for these reasons that Darwin reacted strongly against concepts that bore even a hint of hierarchical, essentialistic, or teleological connotation. For these would have been invariably associated by him, and his contemporaries, with the medieval world view whose death knell he hoped to sound. In the light of this background the renewed use of such terms within recent evolutionary theory, to which the essays in this book bear witness, is a cultural phenomenon of some note. In one way this is no more than a matter of choosing words. But the very fact that such words can now be used without importing associations from which Darwin fled is itself worthy of attention.

There are at least two reasons for this possibility. First, as Campbell notes, our increasingly detailed knowledge of those physical mechanisms whose structures and functions we find it useful to explicate in terms of hierarchical, essentialistic, and teleological models frees these concepts from their former association with transcendentalism. Second, the cultural situation itself has changed markedly since Darwin's time. Despite the persistence of creationist and vitalist talk, serious public discourse in all fields has generally moved toward a comprehensive naturalism. This cultural recession of the imagination of transcendence allows for an expansion of evolution to incorporate many distinct hierarchical levels in a fully natural world. Thus commitment to naturalism need no longer imply the scorched earth of a universal science of particle motion. Nor, therefore, does it any longer need to be protected by an inappropriate reductionism.

These cultural shifts may be seen at work by considering, for example, the historical fate of the "teleomechanist" approach to functional biology recently described by Lenoir (1982). Teleomechanists, who developed the science of embryology in the nineteenth century, hoped to validate empirically organic laws that were not fully reducible to physics or chemistry. Until recently historians of science, dominated by positivist programs, have tended to conflate teleomechanists with the idealist *naturphilosophische* tradition, all the more easily to dismiss them in favor of the full reductionism of Du Bois-Reymond and Helmholtz. Efforts to more faithfully recover this interesting research program are now possible because they can be conducted in a cultural atmosphere that slips between these extreme alternatives. This permits not only historical fidelity but even inspires, as good historical research can, new research programs that bear analogues to teleomeochanist precedents (Bechtel, 1983).

From this point of view, Darwin's conceptual framework, as structured above, can strike us as a well-motivated, highly productive exercise in intellectual asceticism. Under this regime, knowledge has been accumulated that would not be in our possession if, at various points in the history of the research tradition, essentialistic, hierarchical, or anticipatory concepts had been tolerated. Yet our knowledge of evolution now threatens to burst out of this conceptual integument, provoking a conceptual crisis that could force scientists to an unwelcome choice between a fossilized past and an uncertain future. It is fortunate, then, that cultural assumptions and the conditions of public discourse have changed from Darwin's time to such an extent that appeals to reductionism as a protection against transcendence are no longer necessary. If, as we have argued, conceptual assumptions that pervade a culture are in an important sense internal to the development of science, the threat of discontinuity posed by current evolutionary research might be vastly reduced. For knowledge accumulated under a reductionist regimen can now be preserved, expanded, and integrated into a new framework that makes use of conceptual structures which, during the *ancien régime* of teleological creationism, might have been fatal to the successful pursuit of evolutionary knowledge. Indeed, the Darwinian tradition will be seen to have evolved into this new framework, whose lineaments are just now becoming visible. There is no reason to deny, however, that a conceptual revolution will have taken place at the same time. For, as Alisdair MacIntyre has argued,

Every tradition . . . is always in danger of lapsing into incoherence and when a tradition does so lapse it can sometimes only be recovered by a revolutionary reconstitution. . . . It is traditions that are the bearers of reason, and traditions at certain periods actually require and need revolutions for their continuance. (MacIntyre, 1977)

The opposition between evolution and revolution in science can be vastly overestimated. Much of the discussion of scientific revolutions has been conducted in terms of an assumption that revolutions constitute a total break with a previously unchanging status quo. A more useful conception, with quite productive antecedents in social and political thought, is to see revolutions as consolidations of accumulated changes by the introduction of perspectival shifts (Krige, 1980).

This view will, however, be resisted by empiricists and relativists alike. Empiricists will assume that the background assumptions that play such an important role in this conception of theory-change remain external to science. Relativists, on the other hand, will recognize the power of this sort of conceptual shift to shape scientific theories. Nevertheless they will conclude, with their opponents, that when such changes occur, what is taken to be science is not preserved and transformed but destroyed with the passing of one world into another. The difficulties that attend both of these views are by now well known. The fact that the current status of evolutionary theory calls out for an alternative to them is all the more reason to renew our efforts to transcend these fixed positions.

The perspective we have been taking suggests several additional points that may be worth making in conclusion. First, the persistence of natural theological discourse, and perhaps even of residual tendencies to vitalism, together with the desirability of confuting these, might suggest that it is advisable to retain some attachment to reductionistic ideals. These ideals, whether they are given methodological, theoretical, epistemological, or ontological import, are still powerful tools for deconstructing attempts at non-natural explanations. But this may be to take these residual phenomena too seriously, if the effort diminishes our recognition that our knowledge is most perspicuously encoded in a fully naturalistic, but definitely nonreductionistic, conceptual matrix. Such a preoccupation might even retard efforts to attack the most pressing problems in evolutionary biology.

Finally, if our emergent knowledge is in fact best represented by the redeployment of conceptual structures whose utility was rejected at the outset of the Darwinian tradition, the renewed availability of such concepts betokens something of a modest recovery of conceptual tools long forsworn. It is not, of course, the content previously associated with these concepts that is recovered, but their form, which becomes a medium for the articulation of new content. Thus, as Campbell stresses, no return to a premodern world view is implied by the reapplication of such concepts, though perhaps there is here a harbinger of a post-modern world view.

This does mean, however, that the history of science exhibits, in its longest perspective and largest shapes, a pattern of development not unlike that which prevails in the history of other discourses and disciplines. The *novus ordo saeculorum* of eighteenth-century political revolutions was conceived as a renewal of possibilities latent in the distant past, though changed utterly; the most thoroughgoing revolutions in art are, from another perspective, the reconstitution of long-abandoned conventions; and, in general, innovations of the most complete sort derive much of their compelling quality from the fact that they are recoveries of a tradition. What is striking about the history of scientific traditions is not that they depart altogether from this pattern but that they are so successful in building on the past and moving productively into the future. For several centuries this fact has led thinking persons to conclude that scientific inquiry must differ in kind from every other sort of discursive practice. Of late this view has encountered difficulties that have led some to doubt the very fact of scientific success. For our part, we think that this success is unquestionable, and the source of much hope. But this marvel of human understanding comes into view, evoking praise and wonder, just when we see how far it is like, rather than unlike, other forms of discourse.

Note

We thank the following colleagues for their helpful comments on an earlier draft of this essay: Richard Burian, John H. Campbell, C. Dyke, Marjorie Grene, Stuart A. Kauffman, and Merrill Ring.

References

Allen, T. F. H., and T. B. Starr, 1982. *Hierarchy: Perspectives for Ecological Complexity*. Chicago and London: University of Chicago Press.

Arnold, A. J., and K. Fristrup, 1982. The theory of evolution by natural selection: a hierarchical expansion. *Paleobiology* 8:113–129.

Beatty, J. H., 1980. What's wrong with the received view of evolutionary theory? In *PSA 1980*, vol. 2, P. Asquith and R. Giere, eds. East Lansing, Michigan: Philosophy of Science Association, pp. 397–426.

Bechtel, W., 1983. Teleomechanism and the strategy of life. *Nature and System* 5:181–187.

Brandon, R. N., 1978. Adaptation and evolutionary theory. *Stud. Hist. Phil. Sci.* 9:181–206.

Brandon, R. N., 1981. Biological teleology: questions and answers. *Stud. Hist. Phil. Sci.*. 9:181–206.

Brandon, R. N., and R. M. Burian, eds,. 1984. *Genes, Organisms, Populations: Controversies over the Units of Selection*. Cambridge, Massachusetts: The MIT Press. A Bradford book.

Brooks, D. R., and E. O. Wiley, 1984. Evolution as an entropic phenomenon. In *Evolutionary Theory: Paths into The Future*, J. W. Pollard, ed. New York: Wiley, pp. 141–171.

Brown, H., 1977. *Perception, Theory and Commitment: The New Philosophy of Science*. Chicago: University of Chicago Press.

Campbell, J. H., 1982. Autonomy in evolution. In *Perspectives in Evolution*, R. Milkman, ed. Sunderland, Massachusetts: Sinauer Assoc., pp. 190–201.

Campbell, J. H., 1983. Evolving concepts of multigene families. *Isozymes* 10:401–417.

Campbell, J. H., and E. G. Zimmerman, 1982. Automodulation in genes: a mechanism for persisting effects of drugs and hormones in mammals. *Neurobehav. Tox. Terat.* 4:435–439.

Darden, L., and N. Maull, 1977. Interfield theories. *Phil. Sci.* 44:43–64.

Dawkins, R., 1976. *The Selfish Gene*. New York: Oxford University Press.

Dawkins, R., 1982. *The Extended Phenotype*. Oxford and San Francisco: W. H. Freeman.

Dover, G., 1982. Molecular drive: a cohesive mode of species evolution. *Nature* 299:111–117.

Eldredge, N., and S. J. Gould, 1972. Punctuated equilibria: an alternative to phyletic gradualism. In *Models in Paleobiology*, T. J. M. Schopf, ed., San Francisco: W. H. Freeman, pp. 82–115.

Ghiselin, M. T., 1969. *The Triumph of the Darwinian Method*. Berkeley and Los Angeles: University of California Press.

Gould, S. J., 1977. Darwin's delay. In *Ever Since Darwin: Reflections in Natural History*. New York and London: W. W. Norton, pp. 21–27.

Gould, S. J., 1980. Is a new and general theory of evolution emerging? *Paleobiology* 6:119–130.

Gould, S. J., 1982a. Darwinism and the expansion of evolutionary theory. *Science* 216:380–387.

Gould, S. J., 1982b. Change in developmental tuning as a mechanism of macro-evolution. In *Evolution and Development*, J. T. Bonner, ed. Heidelberg: Springer Verlag, pp. 333–346.

Gould, S. J., 1982c. The meaning of punctuated equilibrium and its role in validating a hierarchical approach to macroevolution. In *Perspectives in Evolution*, R. Milkman, ed., Sunderland, Massachusetts: Sinauer Assoc., pp. 83–104.

Gould, S. J., 1983. The hardening of the modern synthesis. In *Dimensions of Darwinism*, M. Grene, ed. Cambridge: Cambridge University Press, pp. 71–93.

Gould, S. J., and N. Eldredge, 1977. Punctuated equilibria: the tempo and mode of evolution reconsidered. *Paleobiology* 3:115–151.

Gould, S. J., and R. C. Lewontin, 1979. The spandrels of San Marco and the panglossian paradigm: a critique of the adaptationist program. *Proc. Roy. Soc. Lon. B*: 205:581–598.

Grene, M., ed., 1983. *Dimensions of Darwinism*. Cambridge: Cambridge University Press.

Guillemin, R., 1978. Peptides in the brain: the new endocrinology of the neuron. *Science* 202:390–402.

Hacking, I., 1983. *Representing and Intervening*. Cambridge: Cambridge University Press.

Hull, D. L., 1976. Are species really individuals? *Systematic Zoology* 25:174–191,

Hull, D. L., 1978. A matter of individuality. *Phil. Sci.* 45:355–360.

Hull, D. L., 1980. Individuality and selection. *Ann. Rev. Ecol. Syst.* 11:311–332.

Hunkapiller, T., H. Huang, L. Hood, and J. H. Campbell, 1982. The impact of modern genetics on evolutionary theory. In *Perspectives in Evolution*, R. Milkman, ed., Sunderland, Massachusetts: Sinauer Assoc., pp. 164–189.

Kimura, M., 1979. The neutral theory of molecular evolution. *Scientific American* 241(5):98–126.

Kimura, M., ed., 1982. *Molecular Evolution, Protein Polymorphism and Neutral Theory*. Tokyo: Japan Scientific Societies Press and Berlin: Springer Verlag.

King, J. L., and T. H. Jukes, Non-Darwinian evolution. *Science* 64:788–798.

Krige, J., 1980. *Science, Revolution and Discontinuity*. Sussex: Harvester Press and Atlantic Highlands, New Jersey: Humanities Press.

Kuhn, T., 1962. *The Structure of Scientific Revolutions*. Chicago: University of Chicago Press.

Kuhn, T., 1970a. *The Structure of Scientific Revolutions*. Second ed. (enlarged). Chicago: University of Chicago Press.

Kuhn, T., 1970b. Reflections on my critics. In *Criticism and the Growth of Knowledge*, I. Lakatos and A. Musgrave, eds. London and New York: Cambridge University Press, pp. 231–278.

Laudan, L., 1977. *Progress and Its Problems: Toward a Theory of Scientific Growth.* Berkeley and Los Angeles: University of California Press.

Lenoir, T., 1982. *Strategy of Life: Teleology and Mechanism in Nineteenth Century Germany Biology.* Dordrecht: Reidel.

Levins, R., 1966. The strategy of model building in population biology. *American Scientist* 54:421–431.

Levins, R., 1968. *Evolution in Changing Environments.* Princeton: Princeton University Press.

Lewontin, R. C., 1974. *The Genetic Basis of Evolutionary Change.* New York: Columbia University Press.

Lewontin, R. C., 1982a. Organism and environment. In *Learning Development and Culture,* H. C. Plotkin, ed. New York: John Wiley and Sons, pp. 151–170.

Lewontin, R. C., 1982b. Review of R. Milkman, ed., *Perspectives on Evolution.* Paleobiology 8:309–313.

MacIntyre, A., 1977. Epistemological crises, dramatic narrative and the philosophy of science. *Monist* 60:453–471. Reprinted in *Paradigms and Revolutions,* G. Gutting, ed. Notre Dame: Notre Dame University Press, pp. 54–74.

Mayr, E., 1942. *Systematics and the Origin of Species.* New York: Columbia University Press.

Mayr, E., 1963. *Animal Species and Evolution.* Cambridge, Massachusetts: Harvard University Press.

Mayr, E., 1982. *The Growth of Biological Thought: Diversity, Evolution and Inheritance.* Cambridge, Massachusetts: Harvard University Press.

Mayr, E., 1983. How to carry out the adaptationist program. *American Naturalist* 121:324–334.

Milkman, R., ed., 1982. *Perspectives on Evolution.* Sunderland, Massachusetts: Sinauer Assoc.

Nagel, E., 1961. *The Structure of Science.* New York: Harcourt, Brace and World.

Nagel, E., 1977. Teleology revisited. *Journal of Philosophy* 74:261–301.

O'Grady, R. T., 1984. Evolutionary theory and teleology. *Journal of Theoretical Biology* 107:563–578.

Ospovat, D., 1981. *The Development of Darwin's Theory.* Cambridge: Cambridge University Press.

Pattee, H., 1970. The problem of biological hierarchy. In *Toward a Theoretical Biology III,* C. H. Waddington, ed. Chicago: Aldine, pp. 117–136.

Pattee, H., ed., 1973. *Hierarchy Theory: The Challenge of Complex Systems.* New York: Braziller.

Popper, K., 1974. Scientific reduction and the essential incompleteness of all science. In *Studies in the Philosophy of Biology,* J. Ayala and T. Dobzhansky, eds. Berkeley and Los Angeles: University of California Press, pp. 259–284.

Prigogine, I., 1980. *From Being to Becoming*. San Francisco: W. H. Freeman.

Prigogine, I., and I. Stengers, 1984, *Order Out of Chaos*. New York: Bantam.

Ruse, M., 1973. *Philosophy of Biology*. London: Hutchinson.

Ruse, M., 1982. *Darwinism Defended: A Guide to the Evolution Controversies*. Reading, Massachusetts: Addison-Wesley Publishing Company.

Sahlins, M., 1976. *The Use and Abuse of Biology: An Anthropological Critique of Sociobiology*. Ann Arbor, Michigan: University of Michigan Press.

Simon, H., 1962. The architecture of complexity. *Proc. Amer. Phil. Soc.* 106:467–482.

Simon, H., 1973. The organization of complex systems. In *Hierarchy Theory*, H. Pattee, ed. New York: Braziller.

Simon, H., 1981. *The Sciences of the Artificial*. Cambridge, Massachusetts: The MIT Press.

Simpson, G., 1944. *Tempo and Mode in Evolution*. New York: Columbia University Press.

Sober, E., and R. C. Lewontin, 1982. Artifact, cause and genic selection. *Phil. Sci.* 49:157–80. Reprinted in *Genes, Organisms, Populations: Controversies over the Units of Selection*, R. Brandon and R. Burian, eds. Cambridge, Massachusetts: The MIT Press. A Bradford book.

Stanley, S., 1975. A theory of evolution above the species level. *Proc. Nat. Acad. Sci.* 72:647–650.

Stanley, S., 1979. *Macroevolution: Pattern and Process*. San Francisco: W. H. Freeman.

Stebbins, G., and F. J. Ayala, 1981. Is a new evolutionary synthesis necessary? *Science* 213:967–971.

Toulmin, S., 1972. *Human Understanding*, vol. 1. Princeton: Princeton University Press.

Wiley, E. O., and D. R. Brooks, 1982. Victims of history: a non-equilibrium approach to evolution. *Systematic Zoology* 31:1–24.

Williams, G. C., 1966. *Adaptation and Natural Selection*. Princeton University Press.

Wilson, A. C., S. Carleson, and T. J. White, 1977. Biochemical evolution. *Ann. Rev. Biochem.* 46:573–639.

Wilson, E. O., 1975. *Sociobiology: The New Synthesis*. Cambridge, Massachusetts: Harvard University Press.

Wimstatt, W. C., 1980. Reductionistic research strategies and their biases in the units of selection controversy. In *Scientific Discovery: Case Studies*, T. Nickles, ed. Dordrecht: Reidel, pp. 213–259. Reprinted (in part) in *Genes, Organisms, Populations: Controversies over the Units of Selection*, R. Brandon and R. Burian, eds. Cambridge, Massachusetts: The MIT Press. A Bradford book.

Notes on the Contributors

Francisco J. Ayala is Professor of Genetics at the University of California, Davis. His most recent book is *Population and Evolutionary Genetics: A Primer* (1982). With Dobzhansky, Stebbins, and Valentine, he wrote the influential text *Evolution* (1977). He edited the interdisciplinary volume *Studies in the Philosophy of Biology* (1974). He is a member of the National Academy of Sciences.

Robert Brandon is Associate Professor of Philosophy at Duke University. He has published a number of studies in the philosophy of biology. He has recently co-edited, with Richard Burian, *Genes, Organisms, Populations: Controversies over the Units of Selection* (1984).

Richard M. Burian is Professor and Head of the Department of Philosophy at Virginia Polytechnic Institute and State University. He has written extensively in philosophy of science and biology. Co-editor with Robert Brandon of *Genes, Organisms, Populations: Controversies over the Units of Selection*, he is currently working on a book on the conceptual history of the gene.

John H. Campbell is Associate Professor of Anatomy at the University of California, Los Angeles. His research interests have been in bacterial and viral evolution and more recently in the evolution of complex genetic systems.

David J. Depew is Professor of Philosophy at California State University, Fullerton. His work centers on the relation between nature and culture in ancient Greek and modern thought, and on problems about the scientific explanation of cultural phenomena. He is the editor of *The Greeks and the Good Life* (1980).

C. Dyke teaches philosophy at Temple University. He is the author of *Philosophy of Economics* (1981) and of the text for the telecourse *Through the Genetic Maze*. He is currently working on a book that attempts to explain the following passage: "Men can be distinguished from animals by consciousness, religion, or anything else you like. They themselves begin to distinguish themselves from animals as they begin to produce their means of subsistence, a step which is conditioned by their physical organization."

Marjorie Grene is Professor of Philosophy, Emerita, at the University of California, Davis. Some of her books are *A Portrait of Aristotle* (1963), *The Knower and the Known* (1969), *Approaches to a Philosophical Biology* (1969), and *The Understanding of Nature: Essays in Philosophy of Biology*. She has edited *Topics in the Philosophy of Biology* (1975) (with E. Mendelsohn) and *Dimensions of Darwinism* (1983). She is a member of the American Academy of Arts and Sciences.

Stuart A. Kauffman, M.D., is Professor of Biochemistry and Biophysics, and of Biology, at the University of Pennsylvania. His areas of research are developmental genetics of *Drosophila*, cell cycles, models of complex genetic networks, and prebiotic evolution.

Ernst Mayr is Alexander Agassiz Professor of Zoology, Emeritus, at the Museum of Comparative Zoology at Harvard University. He has published in ornithology, systematics, evolutionary biology, and in the history and philosophy of biology. Among his publications are *Systematics and the Origin of Species* (1942), *Animal Species and Evolution* (1963), and *The Growth of Biological Thought* (1982).

Gunther S. Stent is Professor of Molecular Biology at the University of California, Berkeley, and a member of the Max Planck Institute for Molecular Genetics. His research interests have been in molecular genetics and more recently in developmental neurobiology. He has also written extensively in the history and philosophy of science. His recent books include *Neurobiology of the Leech* (1981) and *Morality as a Biological Phenomenon* (1981). He is a member of the National Academy of Sciences.

Bruce H. Weber is Professor of Chemistry and Molecular Biology at California State University, Fullerton. His research interests include protein structure-function-evolution and the conceptual development of bioenergetics. He is the author of *Peptide and Amino Acid Analysis* (1975).

Index

For names of authors of cited literature,
refer to the list of references at the end of
each chapter.